Circuit Design for Electronic Instrumentation

Circuit Design for Electronic Instrumentation

Analog and Digital Devices from Sensor to Display

Darold Wobschall

Second Edition

McGraw-Hill Book Company

New York   St. Louis   San Francisco   Auckland   Bogotá
Hamburg   Johannesburg   London   Madrid   Mexico
Milan   Montreal   New Delhi   Panama
Paris   São Paulo   Singapore
Sydney   Tokyo   Toronto

Library of Congress Cataloging-in-Publication Data

Wobschall, Darold.
　Circuit design for electronic instrumentation.

　Includes index.
　1. Electronic instruments—Design and construction.
I. Title.
TK7878.4.W62　1987　　　　621.3815′3　　　86-20866
ISBN 0-07-071231-X

Copyright © 1987, 1979 by McGraw-Hill, Inc. All rights reserved.
Printed in the United States of America. Except as permitted
under the United States Copyright Act of 1976, no part of this
publication may be reproduced or distributed in any form or by
any means, or stored in a data base or retrieval system, without
the prior written permission of the publisher.

1234567890　DOC/DOC　893210987

ISBN 0-07-071231-X

*The editors for this book were Daniel A. Gonneau and Dennis Gleason,
the designer was Naomi Auerbach, and the production supervisor
was Teresa F. Leaden. It was set in Century Schoolbook
by The William Byrd Press, Inc.*

Printed and bound by R. R. Donnelley & Sons Company.

Contents

Part 3   Signal Amplification and Processing

Part 5   Power Circuits

Preface

With the proliferation of electronic devices there has been a corresponding expansion of electronic literature and a subdivision into areas of specialization which treat only certain aspects of instrument design. As a consequence, students and nonspecialists may spend more effort in finding electronic design information than in designing itself or may not even become aware of standard techniques. I have attempted to simplify the design process by gathering and describing standard devices and techniques in a single, reasonably compact book. The designer can then go to a single source for ideas and approaches to an electronic design problem.

The objective of this book is to provide guidance in the design of complete electronic instruments, from input to output or sensor to readout. Therefore interfacing and interrelation of circuits are emphasized. All the parts of a digital thermometer, for example, which include a thermosensor, an analog amplifier, an analog-to-digital converter, and digital displays, are described.

Sections of the text cover temperature, electrooptical, and displacement sensors, analog and digital devices, analog and digital signal processing, digital displays, analog-to-digital conversion, data transmission, and data recording. Many or most instruments in use today utilize several of these circuits or devices.

While microprocessor systems are not covered in full detail, many microprocessor-based instruments employ the circuits described here as peripheral devices. The discussions of multiplexing and data transmission are especially appropriate. The chapter on microprocessors expands on the background by describing the interfacing of microprocessors to external devices.

Coverage of a wide variety of topics in compact form requires brevity and careful selection. Preference is given here to more basic and useful circuits. Sufficient explanation is included to allow the reader to understand circuit operation, but extensive mathematical treatment and detailed descriptions of specialized topics are avoided. Should an interest in understanding the details of specific circuits be kindled, I expect that most readers will prefer, as I do, to consult specialized texts or analyze the circuit themselves.

Students of electronics engineering design will find this book a useful reference text. In one sense, design is not taught by books but through practice and laboratory experience. It must be done by the student. A text can only complement the process by providing guidelines and the basic circuits which are the building blocks of electronic instruments. This text is intended to do so while not relieving the student of exercise by leaving no detail unstated. Students in a senior-level electronic design course, for which this book was developed, design, analyze, construct, and test instruments on an individual basis as a part of the course requirements. Such projects, even in the academic environment, provide experience akin to that encountered industrially and are widely recognized as a valuable facet of an engineer's education. Several examples of how the various circuits are combined to form complete instruments are given in the text. Design problems are also included to provide further design exercise for students.

It is the fate of any electronics book which discusses specific applications and devices to become outdated in parts. In this second edition, circuits have been updated to employ newer and better devices. Another major revision is that the sections on sensors and communications have been enlarged in response to their increasing importance in electronic instrumentation. In keeping with the goal of this book as a compact circuit reference, more basic and less used information on device characteristics has been deleted to make room for the new material.

I am indebted to many students who have used earlier drafts and editions of this book and whose suggestions and encouragement have resulted in the present edition.

I wish to thank my wife, Katrina, for proofreading when she would rather be planting tulips and also for the views of flowers which made the writing of this book a more pleasant task.

Darold Wobschall

Circuit Design for
Electronic
Instrumentation

Semiconductor Devices and Basic Circuits

An Approach to Electronic-Instrument Design

The design of an electronic instrument consists largely of the creative arrangement of proven circuits and devices to achieve a chosen performance goal. Of course, technical requirements must be met in the choice and interconnection of subunits. A textbook cannot teach creativity, but it can describe the devices and circuits which are the ingredients of the design process. It also can provide examples. Actual experience is the best teacher. This chapter has been written as an aid to the inexperienced designer, who may lack confidence or general guidelines to instrument design.

1-1 The Parts of an Electronic Instrument

The purpose of an electronic instrument is typically the measurement and display of a physical parameter. Generally an instrument can be divided into the sections indicated by the block diagram of Fig. 1-1. An input sensor is a device with a measurable electrical parameter which is a function of, and often linearly proportional to, the physical parameter being measured. For example, the resistance change (electrical parameter) of a thermistor (sensor) is proportional to the temperature (physical parameter), at least over a limited range. Usually some type of read-in circuit is present which converts the change in the electrical parameter of the sensor into an electrical signal which can be easily measured. Signal voltage or frequency, in particular, can be measured accurately. A sensor read-in or converter circuit may be quite simple. For the thermistor, a dc voltage source and

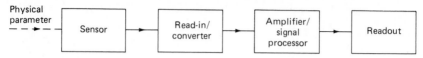

Figure 1-1   General block diagram of an electronic instrument.

a resistor are all that is needed to convert the resistance change into proportional voltage change (signal).

Amplification and perhaps further processing of the signal are often required before it is suitable for readout. Signal processing can be elaborate, involving separation into frequency components, timing of various segments, or conversion from ac into dc. Signal readout may consist of a panel meter, a digital display, or perhaps conversion into a form suitable for transfer to a digital computer. In addition to these parts, an instrument generally requires a power supply, which may be so standard that it needs no separate discussion, and control circuits, which may be an intergral part of the blocks already discussed.

1-2   Circuit Design and Refinement

Instrument design consists in many cases simply of choosing an appropriate circuit for each block from a catalog of available circuits with due consideration to the matching or interfacing of these blocks. It is assumed, of course, that the function, performance characteristics, or specifications are known. A wide variety of sensors and circuits are available. More often than not the problem facing the designer is how to choose compatible circuits from the wide variety available rather than finding a circuit which performs a particular function. Compatibility between sections must be stressed, since without it the required interfacing circuits can increase the complexity of the instrument and may require more effort than a careful selection of basic circuit blocks.

Circuits to perform a particular function can be obtained from a number of sources, e.g., a text like this, journals or magazines concerned with electronic instrumentation, electronic manufacturer's application notes, and local consultants or friends interested in this subject. An experienced designer also builds up a repertoire of circuits or bag of tricks on which to draw, together with a feeling for which circuits, singly or in combination, are easy to implement.

Sometimes a complete and detailed circuit is found to perform the task required, in which case no design is needed; all that remains is construction and test of the instrument. More often than not the circuit found will be incomplete, have too many or too few functions, or otherwise not have quite the right specifications. The designer's task is then to edit, modify, or fill in missing parts of the circuit. Circuits

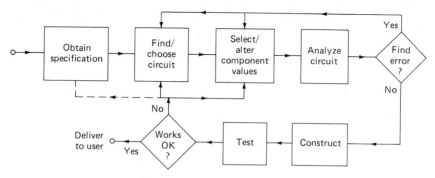

Figure 1-2   Instrument design flowchart.

found in textbooks or application notes generally have one or more components whose values are unspecified, but a formula or procedure is given so that appropriate values to fit a specific application can be chosen. For example, the gain A of a feedback amplifier might be determined by the ratio of two resistors ($A = R_b/R_a$), and the designer is free to set, within limits, the gain by the resistor selection.

Considerable design freedom is usually available, and the student who has been narrowly schooled in the methods of circuit analysis is bothered by this extra freedom at first. A gain of 10 in the example above can be obtained with the combination $R_a = 100$ kΩ and $R_b = 10$ kΩ or $R_a = 20$ kΩ and $R_b = 2$ kΩ, or an infinite number of other choices. Many analytically inclined students would like to see only one solution. They must learn to live with these extra degrees of freedom and even appreciate them.

Once a tentative circuit has been chosen, it should be analyzed, refined, and tested. This is an iterative process, as the flowchart of Fig. 1-2 suggests. In the analysis step, the voltage, current, gain, frequency response, or other appropriate parameters are examined quantitatively. This may only consist of a check on the compatibility of voltages between stages or may involve a more thorough mathematical analysis of the system response. A computer-aided design procedure may be used, especially in complicated high-frequency circuits. If errors or deviations from the given specifications are uncovered, the circuit component values are appropriately modified or if necessary, a better circuit is found. These steps are repeated until no errors are apparent. In other words, the circuit should look good on paper before one proceeds to the next stages, the constructing and testing of the instrument. If problems are uncovered in the testing stage, as is usually the case, the component values may have to be modified, or in the worst case the circuit will have to be scrapped and a new one selected (back to the drawing board).

If the instrument does not meet the specifications provided, the possibility of changing the specifications to fit the actual performance should be considered. Often the specifications provided initially are unnecessarily stringent or unrealistic in some respects, but until the design is complete, the problem is not recognized. Quite possibly a particular requirement was set rather arbitrarily or with a very generous margin of safety which upon closer examination might be modified. Drawing up a complete and realistic set of specifications, in other words, is often part of the design problem and the iterative loop rather than an unalterable input.

1-3 Integrated Circuit Advantages

Instrument design has been enormously simplified by the wide availability of integrated circuits (ICs). Circuits which might take hours to design using discrete transistors take only minutes with ICs. Further, the instrument with ICs will probably work much better, cost much less, and be much easier to construct. Today a beginner can in some cases design and construct an amplifier in an hour which will outperform an amplifier that required days to make by an electrical engineer one or two decades ago. With many IC configurations, the choice of components is simple, even cookbooklike. Novices are not likely to run into many problems with these configurations, and the more experienced designers also appreciate them because they take so little time. Of course, not all circuits involving ICs are easy to design or even understand; there is still an adequate challenge to the experienced engineer. The point is that the easy end of the circuit spectrum is easy indeed and should not be feared by the inexperienced.

An important advantage of most ICs is that they approach ideal behavior closer than the typical discrete circuit. They are more linear and insensitive to power-supply fluctuations, for example. Actually these desirable characteristics are achieved through external negative-feedback stabilization, internal regulation, and stabilization circuits on the IC chip itself, which can be duplicated with discrete transistors.

It would be a mistake to leave the impression that all electronic functions can be carried out best with ICs. An IC may not be available to perform the specialized function the designer has in mind. Some functions are simple enough to be done quite adequately by a single transistor or diode, and an IC is an unnecessary complication. Furthermore, voltage, power, and frequency limitations are generally greater in ICs than in discrete transistors. It will be noted that a number of the circuits described in this text and in commercial instruments involve discrete components.

1-4   Interfacing and Matching Sections

Circuit blocks whick work well individually may not work when connected. Compatibility is a problem in electronic-instrument design, as it is in marriage. Primarily, the interfacing procedure between stages, sections, or blocks involves (1) matching the voltage levels between stages and (2) avoiding excessive loading of a stage by the following stage. In Fig. 1-3, an input-output equivalent circuit for two stages is diagramed. Of interest here is the matching of the output of stage 1 to the input of stage 2, where stage 1 might represent any section of an instrument, including a transducer, and stage 2 the following section. The output of stage 1 is represented by a voltage generator v_o with a series resistance R_o, referred to as the stage output resistance or impedance. This equivalent circuit model for a circuit is rather general and known as the Thevenin equivalent. The input resistance of stage 2 is connected to stage 1, the load on stage 1 is R_i, and a current i_{12} flows from 1 and 2.

At the output terminals of stage 1, the voltage v_i, which is equal to the voltage input of stage 2, is

$$v_i = v_o - R_o i_{12} = \frac{v_o}{1 + R_o/R_i} \tag{1-1}$$

Usually R_o and R_i are not constant or accurately known. Precise predictions of circuit behavior cannot be made unless the effect of these resistances is made negligible. As Eq. (1-1) indicates, this can be done by making the output impedance R_o small and/or the input impedance R_i high, that is $R_o \ll R_i$, in which case $v_i = v_o$ under all conditions. Circuit behavior is often described and derivations made under the implicit assumption that this condition is met. Should this not be the case, an additional high-input, low-output impedance amplifier such as the noninverting operational amplifier (op-amp) configuration (perhaps unity gain) can be added, as described later. Digital circuits present a special problem, which will be discussed in Chap. 5.

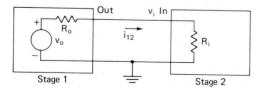

Figure 1-3   Stage input-output equivalent circuit.

1-5 To Build or to Buy?

The question often arises whether it is less costly to design and build an instrument from components than to purchase a complete unit from a commercial source. The production of electronic circuits even in modest quantities results in a much lower unit construction cost. Equally or more important, the cost of design and debugging the instrument can be spread over a large number of units. Of course if no commercial instrument can be found, users have no choice but to build it themselves or have it custom-built. It might appear that the cost advantage lies heavily on the side of the commercial instrument, but this is often not the case. A commercial company must write service manuals, maintain a service group, advertise, and carry administrative overhead. The instrument must also be as versatile as possible to appeal to a reasonably wide range of users. The increased cost of a higher-performance unit is only partly compensated for by the economics of greater production. If the required instrument versatility and performance are much more limited than a commercially available unit, it is quite possible that the user-built unit will be cheaper.

Consider the audio-frequency oscillator as an example. Wide-range, variable-frequency, and amplitude-test oscillators are available from many manufacturers and are competitively priced. Low distortion, good stability, and accuracy are standard. An individual user building an oscillator of similar versatility would find that the cost is much higher than that of the commercial unit. On the other hand, if a fixed-frequency, fixed-amplitude oscillator is all that this is required, the user can build a unit at a much lower cost with an IC. It has the further advantage of small size and easy incoporation into a system, with less risk that connecting wires will come off or that passing knob twiddlers will misadjust the unit.

If a general guideline can be stated, it is that popular, versatile, multicontrol complex instruments are less costly obtained from commercial services,* while special-purpose, fixed- or limited-range simple units can be assembled by the user at lower cost.

1-6 Analog or Digital?

Unmistakably the trend in instrument design is to increase the utilization of digital circuits. The advantages of digital techniques are widely recognized—reliability, flexibility of data reduction, and easy

---

* The prime example is the oscilloscope. The cost advantage is so heavily on the side of the commercial manufacturer that very few oscilloscope circuits are published, although it is one of the most common instruments in use. No one is foolish enough to make an oscilloscope, at least with the thought of saving money.

and accurate readout. Extreme enthusiasts of digital systems often feel that instruments should be as close to 100 percent digital as possible and that the incorporation of any analog circuits represents a deviation from perfection. However, this is an analog world; few physical parameters vary in a discrete fashion. Analog signals are often best processed by analog devices, and it is not obvious that a conversion to digital form is necessarily an improvement. In any case conversion increases instrument cost, complexity, and response time. As new analog and digital circuits are developed, the best circuit for a particular application changes, and it is unwise to assume without a detailed examination that either the analog or digital approach is necessarily superior.

A few generalizations can be made. In instruments or systems with multiple inputs and/or extensive data processing requirements, the digital approach is usually best. Conversion of analog signals from sensors to standard digital form close to the input of a system simplifies processing the data. Transmission of data to a remote point is usually more reliable and accurate in digital form. Readout of a voltage or other variable to an arbitrary high of precision is possible with a multiple-digit display. On the other hand, with a small instrument requiring both analog input and output, the overhead of conversion from analog to digital and the reverse may be costly and unnecessary. Furthermore, digital readouts are not always easiest to read. Maximization of a tuned-circuit voltage, for example, is definitely easier with a standard (analog) voltmeter than with a digital voltmeter. Finally, since most sensors are analog, some degree of analog signal processing is ordinarily required at the input even in a basically digital instrument.

A competent instrument designer, in short, must be familiar with both analog and digital methods and must consider the alternatives fairly before deciding on a specific approach.

Chapter

2

Operational Amplifiers

This chapter discusses the basic operational-amplifier (op-amp) configurations, or building blocks, out of which numerous more complex analog circuits are composed. Other related linear devices are discussed as well. Most involve feedback to carry out their particular function. The characteristics of these feedback circuits depend primarily on the configuration, i.e., how the circuit is connected, and on the resistors and other passive components. Ideally, and to a great extent in practice, these circuits are independent of the particular op-amp characteristics except for the general requirement of high open-loop gain. Some previous acquaintance with feedback is helpful, but a working knowledge can be acquired by studying the first few configurations in detail.

2-1 Characteristics

The voltage output v_o of an ideal differential-input op-amp, symbolized in Fig. 2-1, is proportional to the difference in voltage of two signal sources (v_+ and v_-), as indicated by

$$v_o = A_o(v_+ - v_-) \tag{2-1}$$

where A_o is the open-loop gain. Usually A_o is quite high (10^4 to 10^5), and applications as a linear amplifier will require a feedback circuit, to be discussed at length. Either polarity can be obtained from the output, and the amplifier can be used for dc as well as ac applications. The inputs labeled $+$ and $-$ are the noninverting and inverting inputs, respectively.

A dual power supply (typically ± 15 V) to furnish the voltages $+V_s$ and $-V_s$ is ordinarily required. Note that amplifier output v_o is given with respect to the power-supply common or ground connection. Most

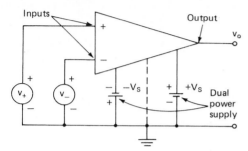

Figure 2-1 Ideal op-amp.

circuit diagrams, including those in this text, do not indicate the dual power supply explicity, but its presence is assumed. Operation with a single-ended supply is possible under conditions discussed elsewhere.

Few electronic devices approach their ideal behavior to the extent that op-amps do, as designers familiar with their application will thankfully acknowledge. Nothing is perfect, however, and several nonideal characteristics must be recognized.

Output saturation

An excessive input voltage will drive the output to positive or negative saturation, a limiting voltage which is somewhat less in magnitude than the power-supply voltage. For many op-amps with a ± 15-V supply, the output will saturate about $+13$ and -13 V. Amplifiers intended for single-supply operation usually have an output voltage range which extends to $-V_s$, which may be grounded. Some lower-power complementary metal-oxide semiconductor (CMOS) op-amps have an output voltage range which fully spans $+V_s$ to $-V_s$. Since the power-supply lines are sometimes called "rails," these amplifiers are said to have a "rail-to-rail" output. If the output is shorted or loaded with an excessively low resistance, the output current may saturate instead. Of course, the maximum output voltage under current saturation is less, possibly much less, than when properly loaded. For the many op-amps, current saturation occurs at 10 to 20 mA.

Offset voltage

The output of a nonideal op-amp will not be zero if the input voltage difference is zero (both inputs shorted to ground) but can be made arbitrarily close to zero by applying a small dc voltage or offset voltage of the proper magnitude and sign to either input. Offset voltage can be a problem when amplifying small dc signals and can usually be neglected at high signal levels. An equivalent circuit which incorpo-

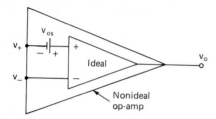

Figure 2-2 Offset-voltage equivalent circuit.

rates the effect of the voltage offset is shown in Fig. 2-2. It utilizes the concept of an ideal op-amp as a subunit of the nonideal amplifier equivalent circuit.

Bias and offset current

As pointed out above, a small bias current (dc) flows out of both inputs of an op-amp to ground. For the usual bipolar junction transitor (BJT) input, the bias current is identical with the base current and is in the range of 0.1 μA. For field-effect transistor (FET) op-amps, input operational amplifiers, the bias current is much smaller, often below 0.1 pA. Bias current is sometimes a problem with high input impedance and/or low-level signal amplification, but it can be ignored in most applications.

Open-loop gain and frequency response

Open-loop gain A_o, as expressed in Eq. (2-1), is a function of frequency, as Fig. 2-3 indicates. Only infrequently must the open-loop gain be known to be better than an order of magnitude, and for many calculations the gain is assumed to be infinite. All op-amps respond to dc gain or zero frequency.

While the high-frequency breakpoints for many operational amplifiers occur at a surprisingly low frequency (5 to 100 Hz), useful gain can be obtained at much higher frequencies. With the unity-gain (feedback) configuration, for example, the high-frequency response extends four to five orders of magnitude higher. The unity-gain (small-signal) frequency response F_U or related gain bandwidth product (GBP) is a useful parameter which can be thought of as the highest usable frequency of the amplifier.

Input (common-mode) voltage range

Op-amps work properly only if input voltages are within a specified range, termed the common-mode operating range. If the input voltage exceeds this range slightly, the output will swing to voltage satura-

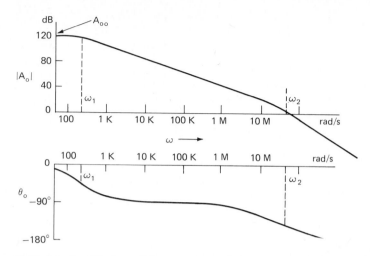

Figure 2-3 Amplitude and phase response (Bode plots) of the open-loop gain of an operational amplifier (LM 356).

tion. If the voltage is exceeded by a large amount, i.e., if the maximum allowable input voltage is exceeded, the device will be destroyed. Newer model ICs have a permissible input voltage range roughly equal to the output-saturation range, for example, ± 14 V with $V_s = \pm 15$ V, so that if the output of one stage is in the linear range, although close to saturation, the next stage has a wide enough common-mode range to follow it. In the case of single-supply-type amplifier, the input range often extends to a value slightly more negative than the negative supply.

Slew rate

Many op-amps cannot deliver anything near full-voltage output at higher frequencies. This limitation is specified by the slew rate, defined as the maximum output voltage change per unit time. If a square wave is applied to the input, the output initially will be a ramp (linear rise) at the slew rate (typically 1 V/μs), even though a more rapid (exponential) rise time would be implied by the frequency response (a small-signal parameter).

Characteristics of some operational amplifiers are given in Table 2-1.

2-2 Noninverting Amplifier

Perhaps the most useful op-amp configuration is the noninverting amplifier (Fig. 2-4). Its purpose is to amplify an input voltage v_i by a

TABLE 2-1. Selected Operational Amplifier Characteristics

No.	Class	V_{os}, mV	$\Delta V_{os}/\Delta T$, μV/°C	I_b, nA	GBW, MHz	Slew rate, V/μs	Min, V	Max, V	$V_s - V_o$, V	$V_o - (-V_s)$, V	Comments
		Offset			Frequency		Supply		V_o range		
LM358	BJT	7	7	100	2	1	3	32	0	2	General purpose
LF351	FET	7	10	0.2	4	13	±5	±18	3	3	General purpose
LM356	FET	2	5	0.05	5	10	±5	±20	3	3	LF357 is uncompensated
741	BJT	5	10	300	1	0.5	±5	±18	3	3	Was industry standard
LM308A	BJT	0.5	5	10	1	1	±5	±15	2	2	Low offset (uncompensated)
TLC251	BiFET	8	2	0.001	0.12	0.04/4	1	16	3	3	Low power (10/100/1000 μA)
LM318	BJT	4	—	0.5	15	70	±5	±20	3	3	Higher frequency
421	BJT	5	15	10	1	0.5	3	30	1.3	1.3	Quad low power, high speed
OP07	BJT	0.01	0.2	3	0.6	0.2	±3	±18	3	3	Precision low offset

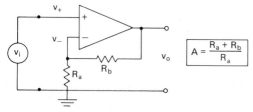

Figure 2-4 Noninverting amplifier.

factor A to give an output voltage of v_o. A fraction β of the output voltage is fed back to the inverting input of the op-amp. Because the sign of the feedback is such as to decrease the magnitude of the output, it is an example of negative feedback.

Calculation of the closed-loop gain ($A = v_o/v_i$) can be made with Eq. (2-1). The voltage of the inverting input is a fraction β of the output

$$v_- = \beta v_o = \frac{R_a v_o}{R_a + R_b} \tag{2-2}$$

where the very small current flow into the input has been neglected. Inserting v_- from Eq. (2-2) into Eq. (2-1) gives an expression for gain

$$A = \frac{v_o}{v_i} = \frac{1}{\beta + 1/A_o} \tag{2-3}$$

For a properly designed amplifier, β is chosen to be at least 100 times greater than $1/A_o$. This usually presents no difficulty because A_o is typically very high (10^4 to 10^6). To a very good approximation, the gain [Eq. (2-3)] becomes

$$A \approx \frac{1}{\beta} = \frac{R_a + R_b}{R_a} \tag{2-4}$$

It should be noted that this expression is independent of A_o, but it is true only if A_o is sufficiently large. At sufficiently high frequencies, this relation will fail because A_o decreases with frequency, but at lower frequencies Eq. (2-4) is quite accurate in practice. Where accurate gains are desired, close-tolerance resistors (1 percent) are used, and closed-loop gains A are limited to 100 or less.

An advantage of the noninverting amplifier is its high input impedance. The minimum gain is unity ($R_a = \infty$, $R_b = 0$), and the maximum useful gain about 10^2 to 10^3. The best range of R_b is 2 to 100 kΩ.

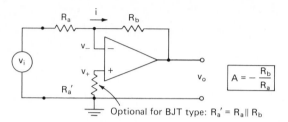

Figure 2-5   Inverting amplifier.

2-3   Inverting Amplifier

A very useful amplifier is the inverting amplifier, shown in Fig. 2-5. The gain can easily be calculated utilizing the infinite-gain approximation ($v_+ = v_-$). The current i, which flows from the input to the output, is

$$i = \frac{v_i}{R_a} = -\frac{v_o}{R_b} \tag{2-5}$$

Therefore, the closed-loop gain A is

$$A = \frac{v_o}{v_i} = -\frac{R_b}{R_a} \tag{2-6}$$

A disadvantage of the inverting amplifier is its relatively low-input impedance, which is equal to R_a because the inverting input is at virtual ground potential. Input impedance, however, is ordinarily much larger than the op-amp output impedance and therefore rarely presents a problem when driven be another op-amp. The input resistance R_a must not be too high (over 100 kΩ) with BJT-type op-amps or the effect of the bias current may become too high. If the bias current is a problem, a bias current compensation resistor ($R_a' \approx R_a$) may be added from the + input to ground. Closed-loop gains A of 0.1 to 100 or 1000 are practical.

2-4   Differential Amplifier

The differential amplifier or subtractor stage has the same properties as the differential-input op-amp except that it employs negative feedback to stabilize the gain A. Following Eq. (2-1), with A replacing A_o, we define the gain as

$$v_o = A(v_1 - v_2) \tag{2-7}$$

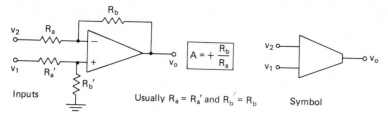

Figure 2-6 Differential amplifier stage.

The simplest of several circuits is shown in Fig. 2-6. It may be thought of as a combination of the inverting amplifier and noninverting amplifier previously discussed, except that the input signal on the noninverting side is slightly reduced by the factor $\alpha = R_b'/(R_b' + R_a')$; that is, $v_+ = \alpha v_1$. The attenuation is required because the gain of the noninverting side is higher.

For many applications, the common-mode rejection ratio (CMRR) = m must be made as large as possible, and in this case R_a' or R_b' may be made variable over a narrow range. To maximize m, the test circuit of Fig. 2-19 can be used.

If the input impedance of this circuit is too low for the desired application, noninverting, perhaps unity-gain, amplifiers are added preceding each input. An improved version of the differential-input amplifier, termed an instrumentation amplifier, is discussed in Sec. 2-10.

2-5 Summing Amplifier

Occasionally the need arises to sum two or more signals. A variation of the inverting amplifier (Fig. 2-7) is well suited for this purpose. Because of the feedback, the summing point v_- is held at zero. As a consequence, the current in any input (i_1, i_2, etc.) is independent of the current in any other, while the output current i is the sum of the

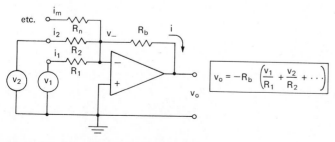

Figure 2-7 Inverting summing amplifier.

inputs. If the resistors are unequal, an extension of Eqs. (2-5) and (2-6) yields

$$v_o = -R_b\left(\frac{v_1}{R_1} + \frac{v_2}{R_2} + \cdots\right)$$

(2-8)

where any number of inputs may be attached.

As with the standard inverting amplifier, signal-voltage sources must have a low impedance, such as provided by the output of a previous feedback amplifier. To correct for the signal inversion, a unit-gain inverting amplifier can be used following the summing amplifier. If a signal is to be subtracted instead of added, a unity-gain inverter can be used before the summing amplifier in line with that signal input.

It should be noted that a two-signal input implifier, with one signal added and the other subtracted, is equivalent to the differential-input stage discussed above, but that amplifier configuration cannot be extended to more inputs because the summing points (v_+ and/or v_-) are not held at ground potential.

2-6   Integrator

The output of an integrator is proportional to the integral over time of its input signal. If the input voltage v_i is constant, the output will be a ramp. In Fig. 2-8 an integrator circuit is given. Calculation of the output voltage as a function of time $v_o(t)$ or, alternatively, its Laplace transform $V_o(s)$ can be made by noting that the input and output currents are equal and that v_- is at virtual ground ($A_o \rightarrow \infty$ approximation)

$$i(t) = -\frac{C\,dv_o}{dt} = \frac{v_i}{R}$$

(2-9a)

$$I(s) = -CsV_o(s) = \frac{V_i(s)}{R}$$

(2-9b)

Rearranging, we obtain

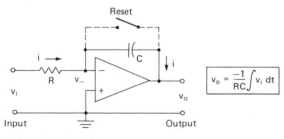

Figure 2-8   Basic integrator.

$$v_o(t) = \frac{-1}{RC} \int_0^t v_i(t') \, dt' + v_{o0} \tag{2-10a}$$

$$V_o(s) = \frac{-V_i(s)}{RCs} \tag{2-10b}$$

where v_{o0} is the output at $t = 0$. Frequently v_{o0} is set equal to zero at $t < 0$ by keeping the capacitor shorted before $t = 0$. The factor $1/RC$ is the equivalent of a gain or sensitivity. Note that the output is inverted and either polarity input is permissible.

A simple integrator will drift into saturation because of the amplifier voltage offset and bias current. FET-type op-amps perform much better than BJT types in this application because of their low bias current. Careful nulling of the offset voltage with attention to the dc component of the input signal is desirable, especially for long integration times. A high-value resistor in parallel with the capacitor will prevent or minimize output drift, but the resulting configuration is more accurately termed an active low-pass filter, which is discussed in Chap. 10.

Capacitor quality is important for integrators. The nonideal capacitor characteristics which degrade integrator performance are (1) leakage resistance, (2) variation of capacitance with temperature, and (3) hysteresis in the charge-discharge cycle. Electrolytic capacitors have the highest leakage,* and one which varies considerably with temperature, age, and average dc voltage across the capacitor (and which must be of one sign since these capacitors are polarized). Ceramic capacitors have a fairly low leakage, but their capacitance is generally very temperature dependent (except "NPO" type). Also they exhibit appreciable hysteresis (Fig. 2-9), expressed as a nonzero or voltage error (V_E) which remains after a capacitor charged and then discharged by the same amount of charge (Q). At the integrator output this corresponds to a nonzero or error voltage (V_E) if exactly equal positive and negative pulses are applied to the integrator input. Mylar or polyester capacitors are much better for integrators since they have low leakage, low temperature coefficients, or low hysteresis. The best capacitors for integrators are polypropylene or polystyrene, but they can be bulky, costly, and not widely available.

A practical integrator is shown in Fig. 2-10. It has an FET which acts as a reset switch and is connected so as to short the capacitor and thus force the output to virtual ground when the FET is turned on (positive gate for an n-channel FET). In some applications a simple

---

* Electrolytic capacitor leakage is often specified as a leakage current (i_L) per microfarad of capacitance (C) per working (maximum rated) volt, V_w. Typically $i_L(\mu A) = 0.02CV_w$ for aluminum electrolytics. Tantalum leakage is a factor of 10 lower.

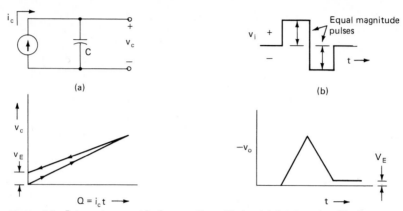

Figure 2-9 Integrator nonideal capacitor effects: (*a*) hysteresis; (*b*) charge-discharge voltage error.

momentary-contact push-button switch will suffice. If a JFET is used, a limiting resistor (typically 100 kΩ to 10 MΩ) must be inserted in series with the gate to prevent excessive drain current when the gate becomes forward-biased. Because $R_{on} \neq 0$, the capacitor will not discharge or reset instantaneously.

In Fig. 2-11, alternate methods of biasing the FET as a switch are presented. Gate voltages sufficient to switch to particular FET on and off (cutoff) should be known. Nearly all available FETs will turn full on and off when driven by an op-amp which switches from positive to negative saturation, assuming a ±15-V supply. Although *n*-channel JFETs turn on for a slightly positive voltage, they may require −2 to −8 V for cutoff. However, because the gate of a JFET (*n*-channel) cannot be biased more than 0.6 V positive without damage and/or breakthrough of the switching signal, it is necessary to limit the gate

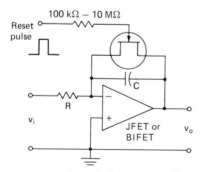

Figure 2-10 Practical integrator. *R* is typically 10 to 100 kΩ (BJT op-amp) or up to 100 MΩ (FET op-amps).

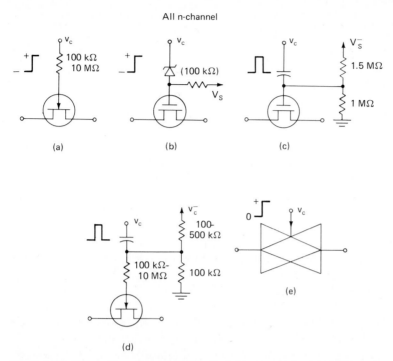

All n-channel

(a)   (b)   (c)

(d)   (e)

Figure 2-11 Alternate *n*-channel FET switch bias methods: (*a*) JFET with limited resistance; (*b*) MOSFET with Zener bias; (*c*) MOSFET biased off, pulsed on momentarily; (*d*) JFET biased off, pulsed on momentarily; (*e*) transmission gate.

current to a low level. A high-value resistor (Fig. 2-11*a*) in series with the gate provides current limit and allows FET turn-on with a positive voltage. The resistor does not interfere with the application of a negative or cutoff voltage since little gate current is drawn.

The differentiator, which is the inverse of the integrator, is discussed in its practical form of a high-pass filter in Chap. 10.

2-7 Current-to-Voltage Converter

A current-to-voltage converter has an output voltage v_o which is proportional to an input current i_i. Signals produced by a very high impedance source, e.g., the photodiode, act as current generators. The circuit is very simple (Fig. 2-12), and again is based on the inverting amplifier. An expression for the output voltage in terms of the input current has already been derived [(Eq. (2.5)].

A simple resistor also acts as a current-to-voltage converter (Ohm's law), but the op-amp circuit reduces the voltage drop at the input v_- to

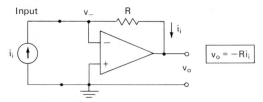

Figure 2-12 Current-to-voltage converter.

nearly zero while producing an easily measurable output voltage v_o. Thus it approaches an ideal milliammeter characteristic of zero voltage drop.

Strictly speaking, the current which flows through the resistor is a sum of i_i and the op-amp bias current, so that the output voltage will have an offset error unless the bias current is compensated or, better, an FET input op-amp is selected.

A voltage-to-current converter has an output current i_o which is proportional to an input voltage v_i and independent of the output load resistance R_L or voltage v_o. In other words, the circuit acts as an ideal current generator. It the load resistance R_L is floating (neither side grounded), the circuit of Fig. 2-13 will work well. Examination of the circuit will reveal that v_- which is equal to v_i by the infinite-gain approximation, is proportional to i_o for any value of R_L and is even independent of v_o, provided v_o is not high enough to cause saturation. This remains true even if R_L is variable or has a quite nonlinear current-voltage relation.

The load resistance R_L can be replaced by reactive load or impedance Z_L, although with some op-amps the circuit will become unstable and oscillate at a high frequency. Should this occur, the oscillation can often be eliminated by altering the external frequency-compensation network, if present, or by adding a small capacitor across the load impedance.

Grounded loads can be driven by the circuit of Fig. 2-14, a variation of the Howland current pump. With the proper choice of resistors, the current into the load i_o becomes proportional to v_i, as seen by the following analysis (again assuming the infinite-gain approximation, $v_- = v_+ = v_o$):

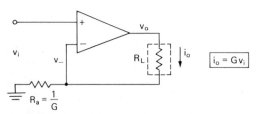

Figure 2-13 Voltage-to-current converter (floating load).

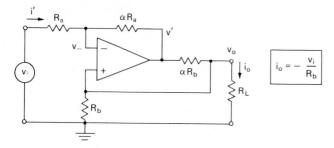

Figure 2-14   Voltage-to-current converter (grounded load).

$$i' = \frac{v_i - v_o}{R_a} = \frac{v_o - v'}{\alpha R_a} \tag{2-11}$$

$$i_o = \frac{v' - v_o}{\alpha R_b} - \frac{v_o}{R_b} \tag{2-12}$$

Note that the resistance ratio α is arbitrary (at least at small signal levels). Solving for v' from Eq. (2-11) and substituting into Eq. (2-12) gives the gain relation

$$i_o = \frac{-v_i}{R_b} \tag{2-13}$$

Output currents of either polarity may be obtained. If the resistors are not accurately balanced, a small output current will be present when $v_i = 0$. It can be minimized by trimming R_a or R_b. Usually α is chosen between 0.2 and 1.0 and R_b is chosen such that v_i is about half the output voltage at saturation when the output current is at maximum.

2-8   Unity-Gain Amplifiers

The unity-gain amplifiers of Fig. 2-15a, b are specialized versions of the noninverting and inverting amplifiers described previously. The main use of the noninverting amplifier is as a high-input-impedance buffer which has an output impedance low enough to drive subsequent stages.

The switchable unity-gain amplifier of Fig. 2-15c can be switched from noninverting to inverting by a digital signal. When the transmission gate (Chap. 3) is turned on, the + input is grounded and thus the circuit acts as a normal inverting amplifier. When the gate is off, no current flows through R' and, again invoking the infinite-gain approximation, this means $v_+ = v_- = v_i = v_o$, and $i = 0$. Thus, the amplifier is nonverting when the gate is turned off.

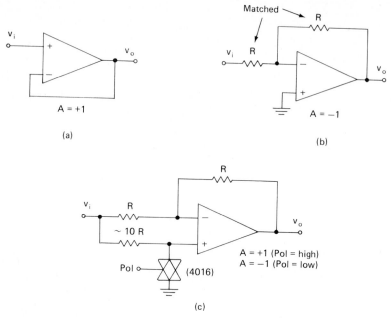

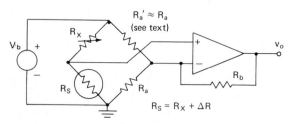

Figure 2-15 Unity-gain amplifiers: (a) noninverting; (b) inverting; and (c) switchable.

2-9 Bridge Amplifier

A convenient op-amp configuration which produces an output voltage proportional to the resistance change ΔR of a resistance sensor is shown in Fig. 2-16. It can be thought of as a combination of an inverting amplifier and a Wheatstone bridge. The two resistors R_a and R_a' not only are two elements of the bridge but also are a part of the feedback resistance. Either an ac or dc bridge voltage supply V_b may be used. Usually the dc supply is obtained from a precision voltage source.

Analysis is simplified by the equivalent circuit of Fig. 2-17, intended to apply when the bridge is near balance ($R_x = R_S$). The terminal

Figure 2-16 Bridge amplifier.

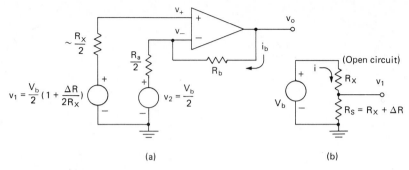

Figure 2-17 Bridge-amplifier equivalent circuit for the (a) noninverting amplifier and (b) sensor-arm subcircuit.

network is replaced by its Thevenin equivalent. At balance, $v_+ = V_b/2$ and at near balance

$$v_1 = \frac{(R_x + \Delta R)V_b}{R_x + (R_x + \Delta R)} = \frac{V_b R_x (1 + \Delta R/R_x)}{2R_x(1 + \Delta R/2R_x)}$$ (2-14a)

$$v_1 \approx \frac{V_b}{2}\left(1 + \frac{\Delta R}{2R_x}\right)$$ (2-14b)

where the binomial expansion for $(1 + x)^{-1}$ with $x = \Delta R/2R_x$ has been used to obtain the last term (terms of the order x^2 are neglected). Because of the high amplifier input impedance, the series impedance of the sensor arm $R_S/2$ is unimportant so that $v_1 = v_+$. Only if the feedback current i_b is zero will $v_2 = v_-$; this is the strict balance condition for which $v_o = V_b/2$. Calculation of the output voltage v_o from the equivalent circuit shows

$$v_o \approx \frac{A\dot{V}_b}{4}\frac{\Delta R}{R_x} \qquad \text{where } A = \frac{R_a + 2R_b}{R_a}$$ (2-15)

Although $v_o = V_b/2$ at strict balance, in practice R_X is usually adjusted so that $v_o = 0$ rather than $R_x = R_S$ as is assumed for Eq. (2-15). Alternatively R'_a can be adjusted so that $v_o = 0$ when $R_x = R_S$ (rather than $R_a = R'_a$).

The bridge amplifier is discussed further in Sec. 4-3.

2-10 Instrumentation Amplifier

An instrumentation amplifier is a fixed-gain differential-input amplifier consisting of three op-amps, as indicated in Fig. 2-18. The gain expression is formally the same as that for an op-amp, i.e.,

$$v_o = A(v_a - v_b)$$ (2-16)

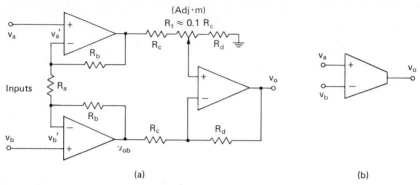

(a)   (b)

Figure 2-18   Instrumentation amplifier: (*a*) circuit; (*b*) symbol.

except that the open-loop gain is replaced by the gain with feedback *A*.

It is basically an improved version of the differential amplifier of Fig. 2-6. Features are (1) high input impedance, especially with FET op-amps on the input; (2) high CMRR, or *m*; and (3) precision high gain. High input impedance is achieved by using the noninverting amplifier configuration on the inputs. Precision high gain is achieved by two stages of feedback amplifiers. High common-mode rejection is achieved by the dual noninverting configuration circuit, which utilizes a common feedback resistor R_a.

The stage gain *A* can be shown to be

$$A = \frac{(2R_b + R_a)\, R_d}{R_a\, R_c} \tag{2-17}$$

The common-mode gain is unity while the differential gain A_d is (R_a + $2R_b$)/R_a. This means that *m* is improved by the factor of A_d by the first stage. The second-stage *m* is determined primarily by the accuracy of the resistor ratio (which in turn determines the gains of the inverting and noninverting sides) and, to a lesser extent, by op-amp nonlinearity. A resistor trim R_t allows maximization of *m* by the test circuit of Fig. 2-19.

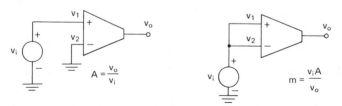

Figure 2-19   CMRR *m* test circuit: (*a*) measurement of voltage gain *A*; (*b*) measurement of common-mode response.

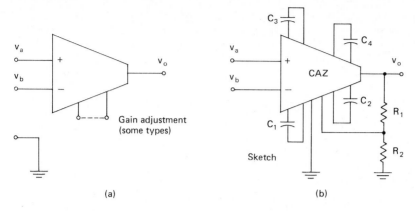

Figure 2-20   IC-type instrumentation amplifiers: (*a*) standard; (*b*) CAZ.

Many users prefer to purchase manufacturer-packaged IC or hybrid instrumentation amplifiers which provide premium performance by careful design and matching (laser-trimming) of components. Most have the same basic configuration as that of Fig. 2-18, and thus the gain can be changed by varying one resistor (R_a), as indicated in Fig. 2-20*a*.

A special type of instrumentation amplifier is the commutating auto-zero (CAZ) amplifier (Fig. 2-20*b*), an IC version of the chopper-stabilized amplifier. Capacitors are switched by analog switches at the commutating or chopping frequency (typically 0.1 to 2 kHz). Each capacitor (C_I) of the differential input section is alternately charged by the input and then applied to the input (v_+) of the following single-ended stage. In this manner the difference voltage between inputs is transferred to a voltage with respect to ground with a very high CMRR. During part of the signal-chopping cycle, the offset voltage is compensated for by charging compensating capacitors to an equal value. When amplifier A is being zeroed, amplifier B is connected to the output, and the reverse.

Major advantages of the CAZ amplifier are the very low offset voltage (<5 μV) and low input current. A disadvantage is the very limited frequency response (<20 Hz) as a consequence of the commutating process. Standard (noninstrumentation) versions of this chopper-stabilized amplifier (e.g., TSC 900) are able to achieve a similar low offset but the normal frequency response of an operation amplifier.

Selected instrumentation amplifier characteristics are listed in Table 2-2. Be aware that occasionally premium-quality operational amplifiers are misleadingly termed instrumentation amplifiers.

TABLE 2-2. Selected Instrumentation Amplifier Characteristics

Type no.	V_{0s}, μV	I_b, nA	Gains	GBW, kHz	CMRR (m), dB	Manufacturer
INA101	0.25	30	1–1000	25	96	Burr-Brown
AMP-01	50	5	Variable	25	90	Precision Monolithic
LM363	100	10	10/100/1000	2	100	National
AD524	50	15	1/10/100/1000	25	120	Analog Devices
4253	1000	0.01	—	5	76	Teledyne

2-11 Transconductance Operational Amplifier and Analog Multiplier

The operational transconductance amplifier (OTA), typified by the CA3080, differs from the standard operational amplifier in two respects. Instead of the output being a voltage source, it is a current source. As a consequence, the gain relation corresponding to Eq. (2-18) becomes

$$i_o = G(v_+ - v_-) \tag{2-18}$$

where G is a tranconductance (gain) with the units of siemens. If a load resistor R_L is added (Fig. 2-21), the circuit acts like a standard op-amp with an output voltage $v_o = R_L i_o$, so that now the performance described by Eq. (2-1) is obtained with $A = GR_L$. Although the open-loop output impedance R_L is higher than in standard amplifiers, it is reduced considerably by the negative feedback employed by most circuit configurations and is not likely to be a problem.

The second functional difference of the OTA is that its gain G is variable. Specifically it is proportional to a set current I_s injected into an input lead.

$$G = \gamma I_s \tag{2-19}$$

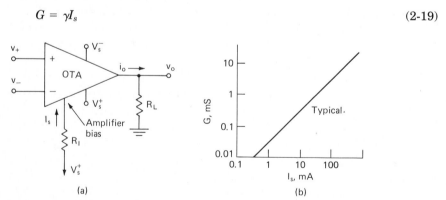

Figure 2-21 Operational transconductance amplifier (OTA): (a) amplifier shown with load resistor and fixed set current I_s; (b) transfer characteristics.

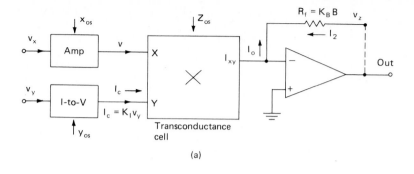

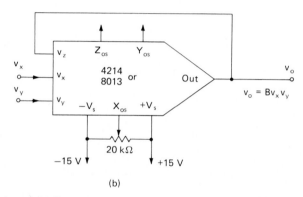

(a)

(b)

Figure 2-22 Analog multiplier.

where γ is a gain-conversion constant for the particular device. This relation is quite linear and holds over a number of decades. The set current should always be positive (inward) and not exceed 0.4 mA. The device is excellent as an amplifier for circuits requiring precise voltage control of gain, as a modulation device, or as an analog multiplier.

An analog multiplier* (Fig. 2-22) provides an output voltage (v_z) proportional to the product of two input voltages v_x and v_y; that is,

$$v_o = Bv_xv_y \tag{2-20}$$

where the gain factor B is usually equal to 0.10 V^{-1}. The input and output voltage range for most four-quadrant multipliers is -10 to $+10$ V. A block diagram of the transconductance multiplier is shown in Fig. 2-23a. It is basically the same as the transconduction operational amplifier, but it is adapted to voltage control for all inputs and can accept both positive and negative voltages on both the x and y inputs

---

* Multiplier circuits are discussed in detail by Y. J. Wong and W. E. Ott, "Function Circuits: Design and Applications," McGraw-Hill, New York, 1976.

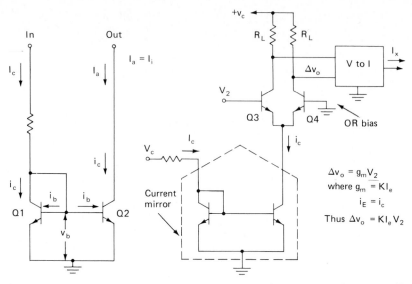

Figure 2-23 Current amplifiers: (*a*) current mirror; (*b*) basic transconductance cell.

(four-quadrant operation). An output or summing amplifier has an extra input, termed the *Z* input, which in the standard multiplication configuration is connected to the output through the internal feedback resistor.

Offset voltages on the *x* and *y*, and to a lesser extent on the *z*, inputs must be carefully compensated unless both input voltages are rather high.

Division is accomplished by incorporating the multiplying circuit into a feedback loop, as indicated in Fig. 2-24. If the v_x and v_y inputs are tied together, the unit can act as a square root circuit.

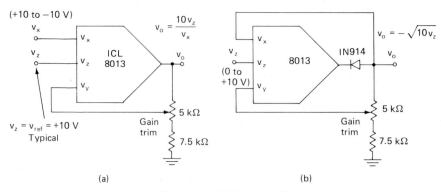

Figure 2-24 Analog division utilizing a multiplier amplifier.

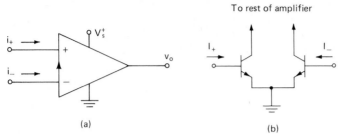

Figure 2-25 Operational current-controlled amplifier (OCA): (*a*) symbol; (*b*) input section.

2-12 Current-Differencing-Amplifier (CDA) Configurations

Another significantly different op-amp is the operational current-controlled amplifier (OCA), also termed a current-difference or Norton amplifier, which has an output voltage proportional to the difference in input currents (i_+ and i_-), that is,

$$v_o = B(i_+ - i_-) \tag{2-21}$$

where B is a gain ($B \approx 3$ GΩ* for the LM3900). Currents must always flow inward. Either dual- or single-ended supply (one side grounded) is permissible, and the most common applications utilize a grounded supply. The inputs are, in fact, transistor bases with emitters connected to the ground (negative supply) lead. Thus the input voltages are 0.4 to 0.6 V more positive than ground (Fig. 2-25). The amplifier symbol is similar to the op-amp symbol with the addition of an arrow on the input to signify current-difference aspect. Maximum output voltage swing is between 0.2 V above ground (most negative) and ($V_s{}^+$ − 1) volts (most positive).

Circuit configurations (Fig. 2-26) based on the current-differencing or OCA type of op-amp are roughly analogous to the ordinary voltage-controlled op-amp described above provided that (1) input voltages are converted into input currents by resistors, and (2) the bias (input) currents are chosen so that the amplifier output does not go to saturation. The circuits here use a single-ended (grounded) supply. If dual-supply operation is desired instead, the V_s terminal replaces ground. For operation in the linear region, the output usually is biased so that its dc (no signal or quiescent) value is somewhat less than half the supply voltage V_s.

The input resistor biasing method can be understood by examining

---

* One siemen or mho is equal to one ohm.

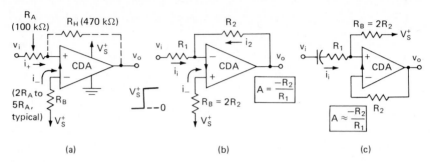

(a)   (b)   (c)

Figure 2-26 OCA configurations: (*a*) comparator; (*b*) inverting amplifier (dc-coupled); (*c*) noninverting amplifier (ac-coupled).

the comparator circuit (Fig. 2-26*a*). Resistor R_B establishes the current threshold by injecting a fixed current $i_- = V_S/R_B$ into the inverting input of the amplifier. The voltage drop between either input and ground is $+0.6$ V and neglected here. When the current into the noniverting input, $i_+ = v_i/R_A$, exceeds i_-, the output will swing to positive saturation (**1**),* while if it is less, the output will be close to ground potential (**0**). Thus the voltage threshold V_T at the input v_i is (neglecting the input voltage drop to ground)

$$V_T = V_S \frac{R_A}{R_B},\qquad\qquad(2\text{-}22)$$

Hysteresis may be added by the positive-feedback resistor R_H. With $+5$-V supply, the output is transistor-trasistor logic (TTL) compatible (see Chap. 3), provided the current difference $i_+ - i_-$ exceeds 10 μA.

It is convenient to analyze OCA circuits by the infinite-gain approximation, which implies that $i_+ = i_-$, provided of course that the amplifier is in the linear operating region. A linear amplifier requires that the output be biased to about half the supply voltage, as indicated above. For the inverting configuration of Fig. 2-26*b*, this is done by choosing R_2 to be half the value of that bias resistor R_B with $i_i = 0$ or

$$\frac{V_S}{R_B} = i_+ = i_- = \frac{v_o}{R_2}\qquad \text{for } i_i = 0 \qquad\qquad(2\text{-}23)$$

which for $R_B = 2R_2$, implies that $v_o(\text{dc}) = V_S/2$. It is assumed that the input v_i has no dc component.

If a signal $i_i = v_i'/R_1$ is now applied, the output voltage change v_o is

---

* The binary digits **0** and **1** for logic 0 and logic 1 are set in boldface to distinguish them from scalars (except in truth tables).

found by the setting $i_i = i_2$ (infinite-gain approximation), and the voltage gain is

$$A = \frac{v'_o}{v_i} = -\frac{R_2}{R_1} \tag{2-24}$$

as expected for the inverting amplifier. This circuit can be converted into an ac amplifier by the addition of an input capacitor. It may be observed that establishment of a bias point is much less precise and more difficult for the CDAs than for the standard op-amps which operate around ground potential.

Numerous circuit configurations can be implemented with the OCA, but they generally do not perform as well as the standard op-amp configurations. Where performance is not exacting, these low-cost configurations can be attractive, especially when single-ended-supply operation is required.

2-13 Voltage-Offset and DC-Bias-Adjustment Methods

Adjustment of the dc level of an op-amp output to zero or a specified bias value can be accomplished by the circuits of Fig. 2-27. The inverting amplifier has the advantage that the offset v_z can be adjusted over the input signal range v_i without affecting the gain. A zero voltage of $-v_z$ for an input of v_i results in an output v_o of $(v_i - v_z)A + v_z$. Only a limited zero range is practical with the noninverting amplifier (Fig. 2-27b), since the resistance in parallel with R_1 must be kept high or the gain will vary with the zero adjustment. An adjustable current flows through R_A to ground in Fig. 2-27b. The maximum current (and thus

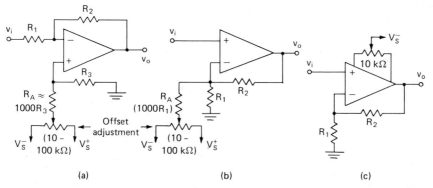

Figure 2-27 Op-amp offset and dc bias circuits: (a) external adjustment for inverting amplifier; (b) external adjustment noninverting amplifier; and (c) internal-offset method.

offset-voltage range) is limited by R_A, which in combination with the input resistor (R_1) acts as a voltage divider.

Some op-amps (Fig. 2-27c) have pins to which an adjustment potentiometer can be connected. The adjustment range of ± 10 mV is adequate to compensate for the input offset but normally is not large enough to bias the output to several volts, as the other circuits shown will do.

If two dc amplifiers are connected in cascade (one driving the next), it is usually necessary to provide only voltage offset to the first, or input, stage since the effect of the offset voltage (and offset compensation) is amplified when applied to the later stage. Offset is usually not needed for ac amplifiers because the undesirable dc component can be taken out by a dc-blocking capacitor or high-pass filter.

2-14 Comparator

The response of an op-amp comparator is full positive output (saturation) when the input difference $v_+ - v_-$ is positive. When the input difference is negative, the output swings to negative saturation. It is similar but not identical to the digital comparator IC (Chap. 3).

Three versions of the comparator are shown in Fig. 2-28. In Fig. 2-28a the threshold voltage V_T is adjustable. Because of the high open-loop gain of the op-amp, an input-voltage rise to only slightly

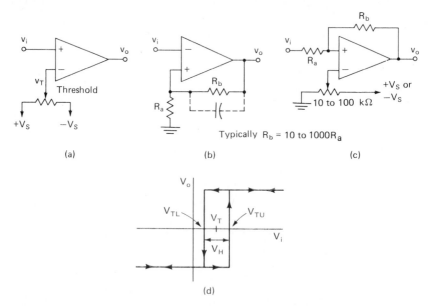

Figure 2-28 Op-amp comparators: (a) simple comparator with variable threshold; (b) inverting comparator with hyteresis; (c) noninverting comparator with hysteresis and variable threshold; (d) response showing hysteresis (V_H).

more positive than V_T will cause the output to swing from negative to positive saturation. The output polarity can be reversed by interchanging the inputs v_+ and v_-.

Positive feedback is utilized in Fig. 2-28b to speed the transition and to provide hysteresis. It can be thought of as a noninverting amplifier with the inputs interchanged. With hysteresis the input voltage required to produce a positive output transition is slightly higher than that required to produce a negative transition (Fig. 2-28d). False transitions due to noisy signals are reduced or eliminated by hysteresis. Hysteresis is increased by decreasing the ratio of R_b to R_a. Typically R_b/R_a is 10 to 1000.

A small capacitor across the feedback resistor will speed the response somewhat, but the slew-rate limitation cannot be overcome and rather sluggish rise times of 2 to 20 μs are typical. If faster response is required, a fast-switching transistor can be connected to the amplifier output or, better, a high-speed digital comparator can be substituted for the op-amp.

2-15 Multivibrator

An op-amp version of the astable multivibrator is described, which is convenient for many control, switching, and timing functions. Digital ICs (Chap. 3) and IC timer circuits (Chap. 16) can perform the same function at higher speed and are ordinarily superior, but in some respects not as versatile. Also the power-supply voltages and signal levels differ from those of most op-amps, an annoyance which can sometimes be avoided by not mixing the two types of ICs.

An astable, or free-running, multivibrator is shown in Fig. 2-29. The output alternately swings from negative to positive saturation with a period determined by the time constant RC and the positive-feedback resistance ratio. If the output v_o is at positive saturation, $v_+ = 0.39v_o$,

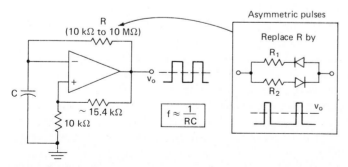

Figure 2-29 Astable multivibrator. For asymmetric pulses, add diode circuit shown.

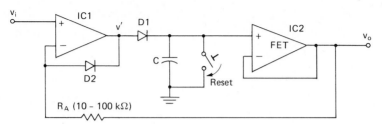

Figure 2-30   Precision positive peak reader.

and until the capacitor charges to a value more positive than this, the output remains at positive saturation. When v_- exceeds V_+, the output swings to negative saturation, and the cycle repeats. Variable frequency can be achieved by making R variable. If an asymmetric wave is desired, the diode-resistor network, which causes the negative and positive charging rates to differ, replaces R.

2-16   Peak Reader

The peak reader shown in Fig. 2-30 utilizes two op-amps. Peak reading is done by charging the capacitor through diode D1. A unity-gain amplifier IC2 follows the capacitor C voltage. Feedback from v_o to the noninverting input of IC1 forces v_o to equal v_i (infinite-gain approximation). Optional diode D2 keeps the IC from saturating during negative-going signals. The feedback resistor R_A limits the current from IC2 during transients. Conversion from a positive to a negative peak reader requires only a reversal of diode polarity.

2-17   Output Limiting and Clipping

Sometimes it is desirable to limit an op-amp circuit output voltage to a point below the saturation value. One reason might be to clip the signal at a precise voltage. The saturation voltage might be used, but it is somewhat variable and dependent on load. Further, limiting circuits keep the op-amp in a linear region, an advantage at higher frequency because the time delays associated with bringing the op-amp out of saturation are avoided and the slew rate limitation is less severe. Two Zener-diode bounding or clipping circuits for the inverting-amplifier configuration are shown in Fig. 2-31. Above the Zener breakdown voltage, the effective feedback resistance R_b, and thus the closed-loop gain, approaches zero. A zero gain at a particular voltage level implies that the output voltage will not rise further as the input rises; i.e., the output signal is clipped. In the circuit in Fig. 2-31a, the negative and positive clipping voltage can be different if Zeners of

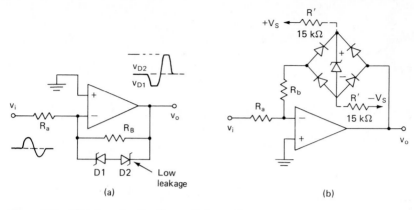

Figure 2-31 Clipping circuits, Zener-diode versions.

different value are selected. The disadvantage of this circuit is the poor frequency reponse and leakage current of Zener diodes.

These disadvantages are overcome by the circuit in Fig. 2-31*b*. The Zener diode always has the same polarity and only the faster bridge switching diodes reverse polarity with an ac signal. The Zener diode bias current from the power supply maintains the Zener at a constant voltage. The Zener resistors R' do not affect the feedback or gain calculation. This is because when the diodes are connected to the output conduct, the resistors in effect appear between the output and power supply as load resistors and not across the feedback resistor R_b.

Chapter

3

Basic Digital Devices

The dominant application of digital ICs is, of course, in computers, but their role in general instrument design is also very important. The most extensive digital series are TTL and CMOS. Several other less common IC series are also useful in instrumentation applications and will be discussed briefly here.

Digital IC devices have two stable states, a higher positive-voltage state termed high or **1*** and a lower voltage state termed low or **0**. This definition applies to conventional binary positive logic. Within a series the voltage range of all devices has the same value and can be interconnected in any order; i.e., the devices are fully compatible. Specification of the logic outputs for various combinations of the inputs is done by means of a truth table and is frequently the only information about a device needed (aside from pin numbers and the voltage levels of the series).

3-1 Gates

Various simple digital gates and the truth tables which specify their function are listed here. The simplest digital device, the digital equivalent of a unity-gain inverter, is the inverter (Fig. 3-1). Its output is a **0** when the input is a **1** and **1** when the input is a **0**.

An OR gate has two or more inputs and one output. Its output will go to **1** if any input is a **1**, as indicated by the truth table in Fig. 3-2a. A NOR gate is the equivalent of an OR gate followed by an inverter (Fig. 3-2b). Note that the small circle on the output side, as elsewhere,

---

* The binary digits **0** and **1** for logic 0 and logic 1 are set in boldface to distinguish them from scalars (except in truth tables).

Truth table

A	Q
0	1
1	0

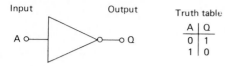

Figure 3-1 Inverter.

indicates inversion. An exclusive-OR (XOR) gate has two or more inputs and one output (Fig. 3-2*c*). As with the regular OR gate, the output will go to **1** if any input is a **1**, except that the output will go to **0** if all inputs go to **1**.

An AND gate has two or more inputs and one output. Its output will go to **1** only if all inputs are **1**, as indicated by the truth table in Fig. 3-3*a*. A NAND (Fig. 3-3*b*) gate is the equivalent of an AND gate followed by an inverter.

Multiple-input gates are also widely available.

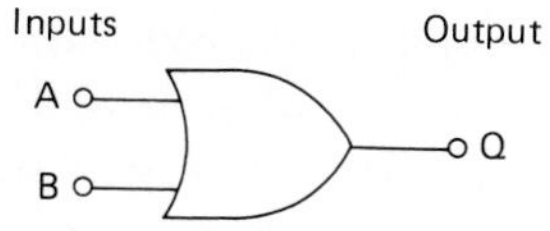

A	B	Q
0	0	0
1	0	1
0	1	1
1	1	1

(a)

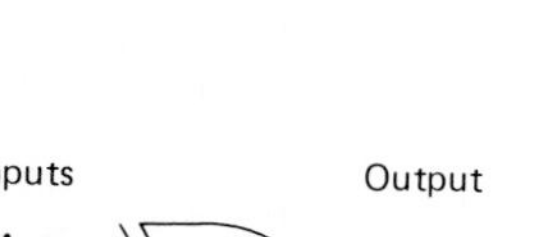

A	B	Q
0	0	1
1	0	0
0	1	0
1	1	0

(b)

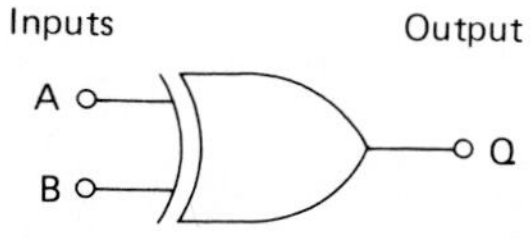

A	B	Q
0	0	0
1	0	1
0	1	1
1	1	0

(c)

Figure 3-2 Two-input (*a*) OR; (*b*) NOR; and (*c*) exclusive-OR (XOR) gates.

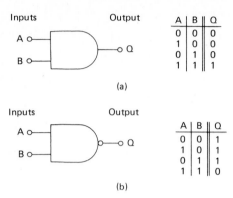

Figure 3-3   Two-input (a) AND; and (b) NAND gates.

A transmission gate (Fig. 3-4) allows bidirectional flow of pulses when on and no flow when off. It is equivalent to a voltage-controlled switch. Although BJT gates are sometimes used, most transmission gates are FET type, commonly CMOS. These devices are symmetric, and either terminal can be connected to the input. Analog as well as digital signals can be switched, but the input and output voltages must fall within a certain range, generally between the most negative terminal V_{SS} and most positive V_{DD}.

A Schmitt trigger (Fig. 3-5) is a logic device with an output which changes state suddenly as a threshold is reached. It is similar to a comparator with hysteresis but is usually operated with logic-level inputs rather than a wide-range analog input. It frequently is employed in analog-to-digital interfacing, especially where a pulse with a slow rise (or fall) is to be changed into a pulse with a fast rise (or fall). Voltage level or time ambiguities, undesirable in digital systems, are reduced or eliminated by the Schmitt trigger.

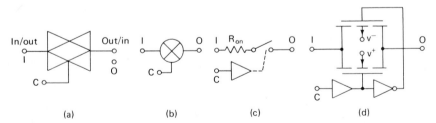

Figure 3-4   Transmission gates: (a) symbol; (b) alternate symbol; (c) electric equivalent; and (d) CMOS analog digital gate.

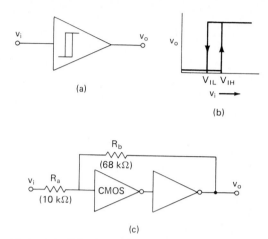

Figure 3-5 Schmitt triggers: (*a*) Schmitt trigger driver symbol; (*b*) input-output transfer function; and (*c*) driver (or two inverters).

3-2 Flip-Flops

An *SR* flip-flop has two inputs and two outputs, as indicated in Fig. 3-6*a*. Application of a pulse to input *S*, that is, momentarily bringing *S* to **1**, will cause the output *Q* to go to **1** until reset by bringing input *R* to **1**. The second output *Q* is the inverse of *Q*. Unlike the gates discussed above, the flip-flop outputs remain in a state indefinitely unless altered by an input pulse, however brief. In a properly designed system both inputs will not be brought to **1** simultaneously, since this produces an indeterminate state.

Often a pair of two-input NOR gates is connected to act as a flip-flop (Fig. 3-6*b*). A NAND-type *SR* flip-flop can be made by substituting NAND gates for the NOR gates, but the input logic sense is inverted.

A *JK*, or master-slave, flip-flop has two logic inputs *J* and *K*, two outputs (Q and $\bar{Q}$), at least two control inputs, a clock CK, and a clear (Fig. 3-7). The *JK* inputs are similar to the *SR* inputs of the *SR* flip-flop, but the outputs do not change until a pulse is applied to the clock input. In digital computers the clock synchronizes the various logic operations.

A key feature of the *JK* flip-flop is that it reverses its output state upon application of a clock pulse if the inputs are both a **1**. In terms of the truth table, the output state Q_n at clock pulse n is changed to Q_{n+1} = $\bar{Q}_n$ at clock pulse $n + 1$, where the bar indicates the inverse state (**1** → **0** and **0** → **1**).

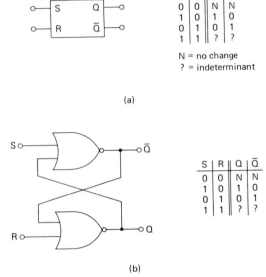

N = no change
? = indeterminant

(a)

S	R	Q	Q̄
0	0	N	N
1	0	1	0
0	1	0	1
1	1	?	?

(b)

Figure 3-6 *SR* flip-flop: (*a*) symbol; and (*b*) dual-NOR circuit.

The type-*D* flip-flop (Fig. 3-8) is similar to the *JK* except it has only one data input *D*. The *D* input is the same as the *J* input of the *JK* flip-flop with the *K* input at *J*. The output changes on the clock rising edge for most types.

A *D* flip-flop can be operated as a binary up counter by typing the *Q* output to the *D* input, since the state to which the *Q* output is to be changed is just the state of the *Q* output. In this case the clock is the pulse input and termed the toggle input.

3-3 Binary Counters

The *JK* flip-flop can be operated as a binary counter because it reverses states upon application of clock pulse when both *JK* inputs are at **1**

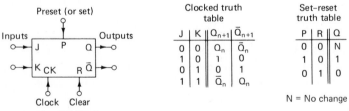

Figure 3-7 *JK* flip-flop.

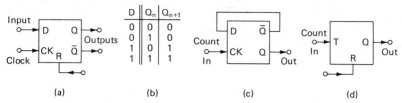

Figure 3-8   Type-D flip-flop: (a) symbol; (b) truth table; (c) connected as divide-by-2 (toggle) counter; and (d) toggle counter symbol.

(Fig. 3-9). Before counting, a pulse is applied to all the clear inputs so that all the outputs (Q_A, etc.) are at **0**. If a series of pulses is applied to the A input (clock), its output will reverse each time the clock pulse goes through one complete cycle, which for many types of flip-flops occurs on the falling edge of the clock pulse.

The output can be interpreted as a binary counter. Decimal equivalents for several binary states are shown in the table. Note that a 4-bit binary counter ranges from 0 to 15 decimal and has an output or carry pulse which can be connected to the toggle input of additional stages. Binary counters are available in many forms (Fig. 3-10), but all have clock and clear inputs.

A binary-coded decimal (BCD) counter is identical to the 4-bit binary counter except for a circuit which resets all the flip-flops to **0** on the tenth count (binary **1010**) rather than the sixteenth. This is done by an internal AND gate with its output connected to clear and its inputs connected to the Q_B and Q_D outputs. Each 4-bit counter is equivalent to one decimal digit and is separately decoded from the binary as a digit from 0 to 9. The decimal digit can be read from a seven-segment

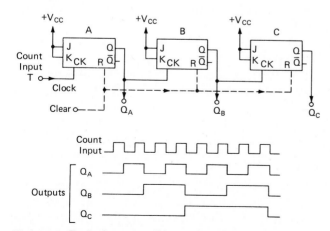

Figure 3-9   Basic three-stage binary counter.

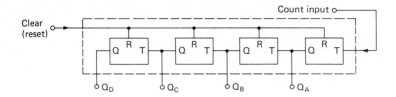

Q_D	Q_C	Q_B	Q_A	Decimal equivalent
0	0	0	0	0
0	0	0	1	1
0	0	1	0	2
0	0	1	1	3
1	0	0	1	9
1	0	1	0	10
1	1	1	1	15

Figure 3-10   Basic 4-bit binary or ripple counter: (*a*) circuit; and (*b*) truth table.

display after decoding (Fig. 3-11). Each segment of the display can be independently illuminated and the decoder/driver selects the proper combinations of segments to form the desired digit, as determined by the BCD input. Detailed decoder display circuits are given in Chap. 5. Counter applications are illustrated in a design example (end of this chapter).

A variety of more complex counters for special applications are available with one or more of the following features: (1) they are presettable to a desired initial value, (2) they can count down as well as up, (3) they can divide by N (a variable), and (4) they can count synchronously. Counters usually have four flip-flops to a package and are encoded in binary or BCD.

By presettability is meant that the counter can be set to any given number (0 to 15 in the case of a 4-bit counter) before counting begins rather than just to zero by the reset input. This process is also referred to as jamming a number into the counter. The number N in binary (or BCD) form must be applied to the data or preset inputs (Fig. 3-12*a*). When the preset-enable line is brought high, the number is transferred to the internal flip-flops and the outputs match the input while the enable is high. When the enable returns to low, normal counting can continue.

An up-down counter can count down as well as up; the direction is controlled by an additional input line (Fig. 3-12*b*). A down counter registering **0000** will change to **1111** on the clock pulse input (or **1001** for BCD). Carry-pulse generation to subsequent stages cannot utilize

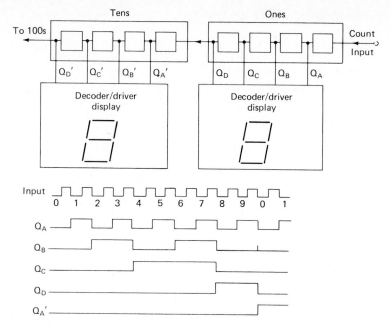

Figure 3-11 Block diagram of two-digit BCD counter and readout.

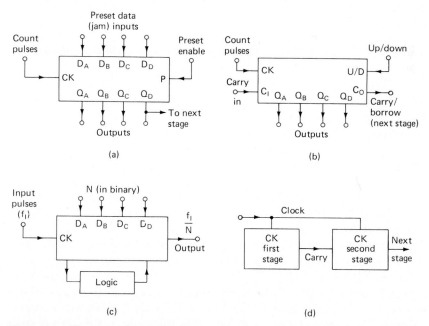

Figure 3-12 Features of special-purpose counters: (*a*) presettability; (*b*) up-down control; (*c*) divide-by-*N*; (*d*) synchronous.

the Q_D output as is done in standard up counters; instead internal logic provides a carry when a zero count is detected.

A divide-by-N frequency counter (Fig. 3-12c) provides one output pulse for each N input pulses, where N is a number in binary form presented to counter data input lines. It functions similarly except that N is a variable, i.e., programmable. Presettable counters can be connected as divide-by-N counters by the addition of external jumpers or logic gates, which depend on the specific device. In this case N (or its complement) in binary form is the preset or jam-input data. Synchronous counters (Fig. 3-12d) are designed to change the output of all stages simultaneously, a desirable feature in applications where timing is critical. In order to avoid propagation delay, the clock must be connected to all stages, but each stage is inhibited unless the carry output of the previous stage occurs.

3-4   Digital Device Characteristics

Complementary metal-oxide-semiconductor (CMOS) IC devices have n-channel and p-channel FETs on the same chip. The simplicity of a CMOS inverter, as pictured in Fig. 3-13a, is notable. When one FET is on, the other is off. In effect, the pair of FETs acts as a single-pole double-throw switch which can connect the output v_o to either supply voltage (**1**) or to ground (**0**) according to the magnitude of the input or gate voltage.

During the switching process, when the gate voltage is intermediate between the **1** and **0** levels, both output FETs (Fig. 3-13a) are partly on. Appreciable current may flow through the transistors at this point, and the instantaneous power dissipation P_D can be quite high, especially at higher supply voltages.

Care must be taken to avoid any open input (gates) on CMOS digital circuits because static charges may build up, causing the device to draw excessive current and overheat. Inputs of all unused sections of an IC package should be tied to ground or $+V_S$. All CMOS inputs have internal diodes on the inputs which are tied to V_{SS} (ground) or V_{DD} ($+V_S$). These diodes help protect against damage by static charges during handling or by excessive input voltages (especially if a series protective resistance is present).

The internal amplifier and output driver of low-power Schottky transistor-transistor logic (LS-TTL) circuits has an input section that can be simplified as indicated in Fig. 3-14a. Only when both inputs are made sufficiently positive, the base voltage will rise to the point where current flows into the base of Q2, the emitter diodes become reverse-biased so that the collector of Q5, which is the output of the IC, is then close to zero (<0.2 V). The overall result is that if either input is low,

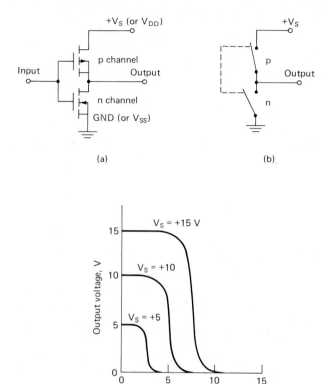

Figure 3-13 CMOS inverter: (*a*) FET schematic; (*b*) output equivalent; and (*c*) transfer characteristics. Input protective diodes to V_{DD} and V_{SS} are not shown.

the output is high, but if both inputs are high, the output is low, verifying the NAND truth table. The static transfer characteristics, which are the dc output voltage as a function of the input voltage, are plotted in Fig. 3-14*b*.

A common mistake made by the novice is to assume that an open input is equivalent to **0**. Actually with TTL it is usually equivalent to **1** because no ground return or sink for the current flowing out of the input is provided. Logic **0** is provided by a grounded input or by a low-value resistor to ground (under 250 Ω for standard TTL or 1 kΩ for LS-TTL). An input connected to the power supply is equivalent to **1**.

Emitter-coupled logic (ECL) is the fastest logic series (0.9 to 2 ns). Speed is achieved not only by emitter coupling between stages but also by operating with small logic-level swings (0.8 V) and with high BJT

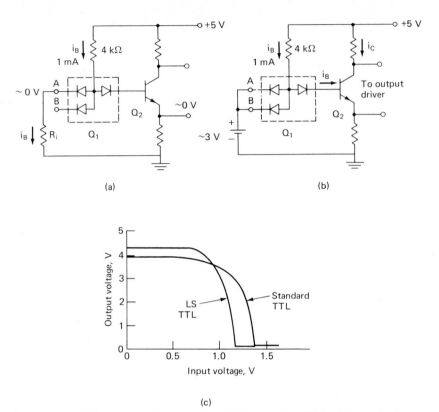

Figure 3-14   TTL NAND gate input section: (*a*) input *A* low; (*b*) both inputs high; and (*c*) transfer characteristics.

emitter currents (Fig. 3-15). High currents imply high power dissipations, and therefore cooling the devices can be a major problem, especially with the highest-speed ECL series (100k). ECL devices are much harder to use than those of other series, especially at maximum speeds.

The power supply for TTL devices V_{CC} is +5.0 V and must be held within 5 to 10 percent. The power supply for CMOS V_{DD} may typically be anywhere between +3 and +15 or 20 V. For CMOS, the supply current is very small, about 10 pA per gate, but it jumps to around 1 mA per gate during the brief time (0.1 μs) required to change states; for TTL, it is 2 to 5 mA per gate. Characteristics of several digital logic series are compared in Table 3-1.

Simple AND and OR gates can be made from diodes (Fig. 3-16). If a positive voltage of sufficient magnitude is applied to either input of the OR gate, the diode will conduct and the output will rise, i. e., go to **1**. The value of the resistor R is unimportant except for loading of the input or the output impedances of the driving or following stages.

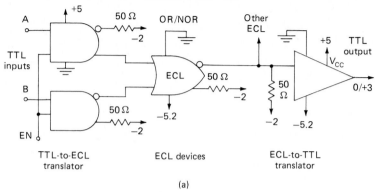

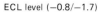

ECL level (−0.8/−1.7)

TTL-to-ECL translator   ECL devices   ECL-to-TTL translator

(a)

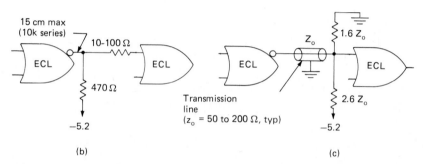

(b)   (c)

Figure 3-15 TTL-to-ECL interfaces: (*a*) TTL to ECL and back; (*b*) interfacing between ECL devices (short wires); and (*c*) ECL interfacing (longer distances).

TABLE 3-1   **Characteristics of Several Logic Series**

Series	Speed Prop. decay, ms	Clock freq., MHz	Power per gate, mW	Supply $+V_S$	Logic level 0	Logic level 1	Example IC no.
CMOS	100*	2*	0.001†	+3 to +15	0	$+V_s$	4011 or 74C02
H-CMOS	10	40	0.001†	+3 to +6	0	$+V_s$	74HC02
TTL	20	15	10	+5	0	+3	7402
LS-TTL	100	30	2	+5	0	+3	74LS02
S-TTL	4	40	18	+5	0	+3	74S02
AS-TTL	2.5	40	4	+5	0	+3	74F02
ECL (10k)	2	180	24	−5.2	−1.72	−0.88	10101‡
ECL (100k)	0.8	400	40	−4.5	—	—	100101‡

* CMOS speed given for simple gates with +5 V supply—higher for complex gates, lower for higher supply voltage.
† Static (higher value at 1-MHz switching speed).
‡ Recommended for those experienced in high-frequency design.

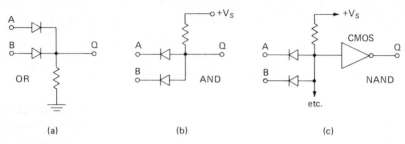

Figure 3-16 Diode gates: (*a*) two-input OR; (*b*) two-input AND; (*c*) multiple-input NAND utilizing CMOS inverter.

A drawback of the diode gate is that the output-voltage change will be less than the input-voltage change because of the diode drop (0.6 V). Further, the rise time of the output is less than that of the input.

For must instrumentation and industrial applications, standard CMOS is usually the best choice because of its low power requirements and high noise immunity. At higher speeds, LS-TTL or high-speed CMOS (HC) may be required. If highest speeds are necessary, advanced low-power- or fast-schottky-type TTL devices (ASL-TTL/F-TTL) or ECL must be used even though implementation is more difficult than with the other series.

3-5   Digital Comparators

A digital comparator is an interface device between analog and digital circuits. The input section is similar to an op-amp, while the output section can be connected directly to TTL or CMOS logic devices (Fig. 3-17). The output transistor collector often must be connected through a load resistor R_L to the digital supply $+V_S$ so that the logic level matches the digital devices to be driven.

3-6   Negative-Logic Notation

Standard positive-logic notation, as used throughout this book, refers to 0 V (approximately) as the low logic level and also as logic 0, or **0**. Correspondingly, the more positive voltage, for example, $+3$ to $+5$ V, is termed high and logic 1, or **1**. From the boolean algebra or mathematical logic point of view, the assignment of **0** or **1** to one of the two stable states of a particular logic device is, of course, arbitrary. Actually, two conventions are embedded in this notation; the assignment of high and low states to the device voltage states and the assignment of **1** or **0** to the high state. Nearly always the more positive state is termed high even if (as for some ECL devices) both logic states may involve negative voltage levels. However, it is not uncommon to designate **0** as the high state, a convention termed negative logic. Designers often find that mixing negative and positive logic throughout a system reduces the number of gates required or is otherwise more compatible with available hardware. To handle the notational problem, two methods are commonly employed: (1) positive logic is assumed for a signal, for example, A, and the bar added, for example, $\bar{A}$, whenever the signal is inverted or (2) alternate symbols are employed for inverting logic devices, including NAND or NOR gates.

The relation of the inverting symbol (small circle) or negative logic is indicated in Fig. 3-18. Here A represents a positive logic signal (**1** = high for A) while $\bar{A}$ is the inverted signal and a is the same signal represented by negative logic (**1** = low for a). In all three cases in the example shown, the effective signal inside the device is low when A is high. In Fig. 3-18b this occurs because the applied signal is inverted by an inverter (circle) present on the input. Figure 3-17c shows the same device as Fig. 3-17a, but a negative-logic symbol a is used. Confusion sometimes occurs in relating the truth table to device operation if the

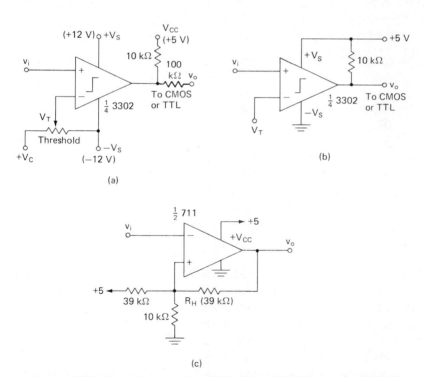

Figure 3-17   Digital comparators: (a) example of use with triple supply; (b) single supply grounded, for which the threshold must be positive; (c) comparator with hysteresis.

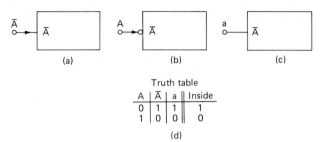

Truth table

A	Ā	a	Inside
0	1	1	1
1	0	0	0

(d)

Figure 3-18 Positive- and negative-logic input symbols.

truth table lists the signal as A (positive logic) when actually $\bar{A}$ must be applied to the input, as in the example shown.

To see the point in the alternative logic symbols for the NOR and NAND gates of Fig. 3-19, note that a standard positive-logic NOR gate used with negative logic acts as a NAND gate (see truth table). Likewise a standard positive-logic NAND gate functions as a negative-logic NOR. These boolean algebra relations are a consequence of De Morgan's theorem. Thus by redrawing the device logic symbols, the logical function in the circuit is emphasized. Negative-logic gate symbols are usually not given in specification sheets, and the equivalent positive-logic device must be selected.

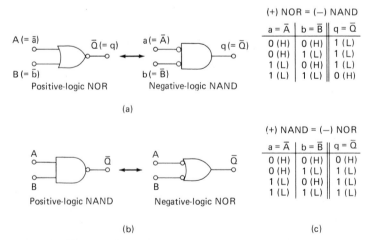

(+) NOR = (−) NAND

a = Ā	b = B̄	q = Q̄
0 (H)	0 (H)	1 (L)
0 (H)	1 (L)	1 (L)
1 (L)	0 (H)	1 (L)
1 (L)	1 (L)	0 (H)

(a)

(+) NAND = (−) NOR

a = Ā	b = B̄	q = Q̄
0 (H)	0 (H)	0 (H)
0 (H)	1 (L)	1 (L)
1 (L)	0 (H)	1 (L)
1 (L)	1 (L)	1 (L)

(b)   (c)

Figure 3-19 Negative-logic gate symbols: (a) NOR as a negative logic AND (b) NAND as negative-logic OR; (c) truth table shows interconversion.

BASIC CIRCUIT DESIGN EXAMPLES

Example 1 Inverting Amplifier (High Z Input)

DESCRIPTION: The circuit provides voltage amplification from dc to 50 kHz.

SPECIFICATIONS
  Gain: -10 and -100 (switch selectable)
  Frequency response: dc to 50 kHz (minimum)
  Input impedance: 10 MΩ (minimum)
  Output impedance: 10 Ω (maximum)
  Output voltage and current: ±10 V at 10 mA up to 5 kHz (minimum) Zero
  offset voltage at output: 0.1 V (maximum)

DESIGN CONSIDERATIONS (Fig. E-1): The output voltage, current, and impedance are within the capabilities of the standard op-amps (with a ±13-V supply). However, an input impedance of 10 MΩ is difficult to achieve with a BJT op-amp, primarily because of the bias current, thus indicating that an FET op-amp is needed on the input.

Two stages are desirable because the gain of 100 at a bandwidth of 50 kHz can be obtained with a single stage only with a premium-quality amplifier (318), while two standard amplifiers in cascade can accomplish the same at much lower cost. Another reason for two stages is that the input impedance of the required inverting amplifier is inherently low, suggesting that a noninverting input stage which does have a high impedance would be needed in any case.

At a gain of 100 the input offset voltage must be within ±1 mV in order to achieve an output offset to ±0.1 V. A trimming potentiometer is needed with standard op-amps to achieve this offset. If a gain bandwidth product of 4 MHz is assumed for both amplifiers, the frequency response is dc to about 200 kHz (at a gain of 100).

Example 2 Decade Counter and Display

DESCRIPTION: The decade counter counts input pulses from an external source. Output is a seven-segment light-emitting diode (LED) display. TTL and CMOS logic versions are given.

SPECIFICATIONS
  Input voltage: CMOS-compatible ($0 \approx 0$ V, $1 \approx 5$ V)
              TTL-compatible [$0 \approx 0.3$ V, $1 \approx 3.6$ V sink 1.6
              mA (0.4 mA, LS type)]
  Digits: Two decimal digits shown (expandable to three or more)

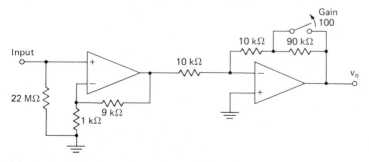

Figure E-1 A two-stage inverting amplifier.

DESIGN CONFIGURATIONS: Interfacing between sections is simplified by selection of one series, either CMOS or LS-TTL. To clear the counter, a momentary **1** level is applied to the clear inputs.

A standard seven-segment common-anode LED driver and display are used (see Chap. 5). Segment resistors limit the current to about 20 mA per segment (140 mA display total).

The CMOS version (Fig. E-2a) uses *a* half of a dual decade counter (4518). As with any CMOS device, all unused inputs (not output) should be grounded (or

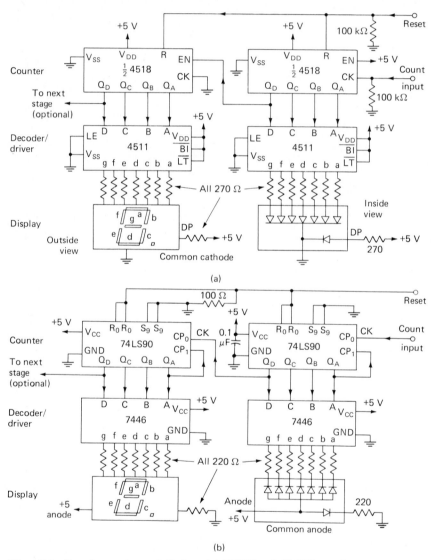

Figure E-2 Decade counter and display: (*a*) CMOS and (*b*) TTL version.

$+V_S$) to avoid excessive power dissipation and random switching of open inputs. A supply between about $+4$ and $+15$ V is acceptable, but the LED segment resistors were chosen for a $+3$-V supply (10 mA per segment). Supply current is largely that delivered to the display.

The LS-TTL counter (Fig. E-2b) requires a jumper wire from the first-stage output Q_A to the second input B. Note that the 74LS90 has two set 9 inputs labeled S_9 as well as two reset 0, or clear, lines. A 1 to either reset overrides the counter pulse input. If reset inputs are left open (ungrounded), the counter will not operate, since with any TTL device an open input is equivalent to 1. Care must be taken to keep the supply between $+4.5$ and $+5.5$ V (at 250 mA) at all times. To avoid supply transients and likely false counts, mount the supply bypass capacitor near the counter.

BASIC CIRCUIT DESIGN PROBLEMS

1 Design a noninverting amplifier with a gain of 500 and a frequency response of dc to 2 kHz. What changes would be required if the frequency response were dc to 50 kHz?

2 Design a peak reader to respond to an input pulse with a voltage v_i range of 0.1 to 1.0 V. The input impedance must be 100 kΩ or over, and the output should drive a standard voltmeter (10 V full-scale, basic movement of 1 mA).

3 Suggest a circuit for a threshold detector which will turn on a lamp if an input voltage exceeds a fixed level (1.0 V). The 6-V lamp draws 100 mA. Note that the lamp exceeds standard op-amp capacities but not those of discrete transistors.

4 Suggest a circuit for an amplifier with a gain of 5 and bias voltage adjustable from at least $+6$ to -6 V.

5 Design an amplifier that will automatically change gain from 100 to 10 if the input signal rises above $+0.1$ V. Include hysteresis so that the gain will not change back (from 10 to 100) until the signal level drops below $+0.02$ V. Indicate the range by a lamp (LED). *Hint*: Gain resistors can be switched in or out by an FET or a transmission gate.

6 Design a peak-to-peak reader that will produce an output v_o equal to the difference between the positive and negative peaks of an input signal. The output voltage v_o must be referenced to ground (as the input voltage is). Peak voltage should be held until reset by a push button.

7 Design a logic device which will turn on an indicator lamp if an input signal A lies between $+1.0$ and $+1.1$ V provided control signal B is negative and signal C is positive. *Hint*: Use comparators and logic gates.

8 Design a frequency divider with outputs of $f/10$, $f/20$, and $f/100$, where f is the input frequency.

9   Design a pulse synchronizer which will change output Q_o state (0 to 1) on the negative-going edge of an input clock pulse (CK) if an enable pulse E is present (E must be at 1 on the falling edge of the clock pulse). The output, once triggered, should remain a 1 until E returns to 0 (more precisely, $Q_o = 0$ occurs when $E = 0$ and CK $= 0$).

Sensors

4

Temperature Sensors

Nearly every electrical property of a material or device varies as a function of temperature and could in principle be employed as a temperature sensor. The requirements of operation over a wide temperature range with high sensitivity, reproducibility, and linearity greatly limit the possibilities, especially if cost, size, and ease of readout are also considered. No sensor available has all these desirable properties but may have the proper combination for a particular application.

4-1 Thermistors

One of the most popular thermosensors is the thermistor, basically a resistor with a high temperature coefficient. A thermistor is a semiconductor in various geometrical configurations to which two leads are attached (Fig. 4-1). One of the advantages of the thermistor is the wide variety of shapes and sizes (down to microscopic) possible. Often a coating is applied for electrical insulation and mechanical protection.

Like most other semiconductors, the standard thermistor has a temperature coefficient α which is negative (resistance decreases with increasing temperature) and quite high, but is nonlinear in resistance-temperature variation. Over most of its range the resistance (R_T) decreases exponentially with temperature (Fig. 4-2)

$$R_T = R_A e^{\beta/T} \tag{4-1}$$

where T is absolute temperature in Kelvin and B and R_A are constants for the device. Most manufacturers express the temperature depen-

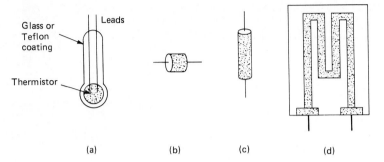

Figure 4-1 Thermistor configurations: (a) coated-bead; (b) disk; (c) diode case; and (d) thin-film.

dence in terms of the resistance (R_a at a specified temperature, e.g., 25°C) and temperature coefficient α, defined as

$$\alpha = \frac{\Delta R_T / R_T}{\Delta T} \tag{4-2}$$

Note that α is the same at all points along the curve, a characteristic of an exponential dependence. Thermistors with a wide range of resistances are available, and for most commercial units the temperature coefficients are in the range of 3 to 5 percent per degree Celsius. A conversion chart (Table 4-1) or graph line (Fig. 4-2) is generally supplied by the manufacturer. For precision units R_o is closely controlled (say ±1 percent), but for standard units only the temperature coefficient is controlled while R_o may vary by 10 percent or more from unit to unit. Thus at most each thermistor requires an individual calibration at one temperature. It is fair to observe that most thermistors are not highly precise sensors because the resistance-temperature

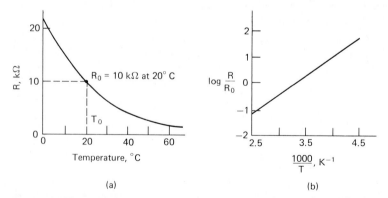

Figure 4-2 Thermistor resistance versus temperature (typical): (a) linear scale and (b) logarithmic plot.

TABLE 4-1   Thermistor Resistance as a Function of Temperature (R_T, = 5 kΩ at 25°C)

Temperature, °C	R_T, Ω	Temperature, °C	R_T, Ω
0	14129	55	1728
5	11335	60	1472
10	9152	65	1260
15	7437	70	1082
20	6080	75	934
25	5000	80	808
30	4134	85	703
35	3438	90	613
40	2873	95	537
45	2414	100	471
50	2038	105	415

characteristics are not highly reproducible. Also, their nonlinearity is a nuisance. Their low cost, high sensitivity, ease of readout, and small size outweigh these disadvantages for many applications.

Two additional parameters can be important, the time constant τ_T and the power dissipation constant η. Internal heating or power dissipation is caused by the voltage applied to the thermistor from the readout or converter unit. Both depend on the device thermal insulation and degree of external heat dissipation; these may be 3 to 10 times different for a thermistor in still air compared with a stirred liquid. Typical values in a liquid are $\tau_T = 0.5$ to 5 s and $\eta = 0.3$ to 3°C/mW.

A simple readout is the voltage divider (Fig. 4-3a), which has an output V_T of

$$V_T = \frac{V_S R_T}{R_T + R_S} \approx \left(\frac{V_S}{R_S}\right) R_T \tag{4-3}$$

High sensitivity is achieved by increasing V_S and V_T, but care must be taken to limit the internal heating effect due to the power dissipation in the thermistor ($P_D = V^2_T/R_T$). Typically a temperature rise of 0.2°C can be tolerated; if we assume $\eta = 0.3$ mW/°C and $R_T = 10$ kΩ, the maximum value of P_D is 6 mW or V_T 7.5 V (maximum).

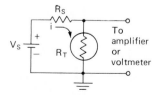

Figure 4-3 Thermistor read-out: simple voltage divider.

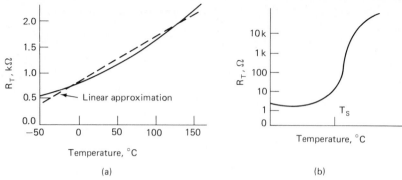

Figure 4-4 Positive-temperature-coefficient thermistors: (*a*) linear and (*b*) switching.

Positive-temperature-coefficient semiconductors, a nonstandard type of thermistor, are available but not popular. Heavily doped silicon (Fig. 4-4*a*) has a fairly linear response. Another type (Fig. 4-4*b*) used for switching exhibits a rapid rise above a specified temperature.

4-2   Readout and Linearization of Resistance-Type Sensors

The resistance-to-voltage converter, suitable for the thermistor or any wide-range resistance sensor, is shown in Fig. 4-5. Conversion is done by the first-stage inverting amplifier with an output of

$$-v_T = \left(\frac{V_{ref}}{R_1}\right) R_T \tag{4-4}$$

If the high accuracy is required, V_{ref} should be a precision-regulated

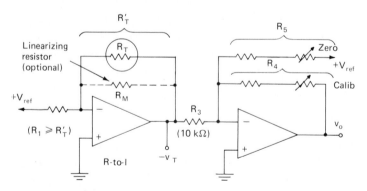

Figure 4-5 Resistance-to-voltage converter for thermistor or other resistance-type sensor.

source (Sec. 10-8), although for most thermistor readout applications a standard power-supply regulator has adequate stability (0.5 percent is typical). Resistor R_1 (with V_{ref}) is chosen such that the thermistor voltage (v_T) produces an acceptable internal temperature rise at the lowest temperature (max R_T). Zero offset and gain are set by the optional second inverting amplifier. The output v_o is proportional to temperature with any desired offset.

$$v_o = \frac{V_{ref}}{R_1} \left(\frac{R_4}{R_3} \right) R_T - V_Z \tag{4-5}$$

where the zeroing voltage is $V_Z = V_{ref}(R_4/R_5)$. Note that the zero (offset) and calibration (gain) controls are noninteracting, a desirable feature. Some users may prefer to digitize the output of the first stage (v_T) directly (possibly after changing v_T to positive polarity by making V_{ref} negative). Offsets for an individual sensor correction can then be corrected by software, a possible nuisance if sensors are exchanged. The R-to-V converter may be used for any resistance-type sensor, including those that require an ac drive, by replacing V_{ref} by an ac source.

The nonlinear (exponential) response of thermistors can be compensated by several methods. One method is to employ a nonlinear amplifier with a response characteristic which is the inverse of the thermistor response.

A simple and surprisingly effective method of thermistor linearization is to add a fixed resistor (R_M) in parallel which is approximately equal to the thermistor resistance at the middle of the temperature range (T_M). The circuit (Fig. 4-6) acts as a thermistor R'_T with effectively half the resistance and temperature coefficient of the actual thermistor (Eq. 4-6) but with a substantial linear range (typically a span of 20 to 50°C).

$$R'_T = \frac{R_M}{2} \left[1 + \frac{\alpha}{2} (T - T_m) \right] \tag{4-6}$$

The compensation technique is based on the observation that the net resistance of a variable resistor in parallel with a fixed resistor R_M changes nonlinearly with the resistance of the variable resistor. By proper choice of a shunt resistance, this nonlinearity compensates the thermistor nonlinearity over a certain range. This can be shown mathematically by expanding the thermistor response R_T as a power series in $\Delta T = T - T_M$ and retaining terms in $(\Delta T)^2$. This linearization technique is not restricted to thermistors nor even to an exponential resistance variation. Several other simple techniques (e.g., series

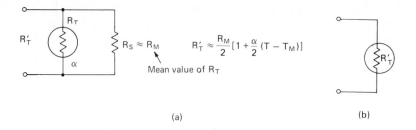

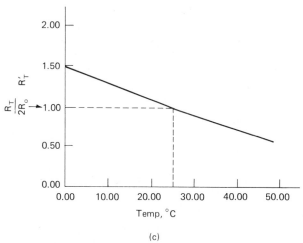

(c)

Figure 4-6 Thermistor linearization by shunt resistance: (a) actual circuit; (b) equivalent circuit; and (c) linearized response.

resistor) are equally, but not more, effective. In Fig. 4-6c, the nearly straight line is the linearized thermistor response. Nonlinearity (least-squares fit) is about 1 percent for 30°C range, better over a narrower range, and much worse over a wider range.

Two (or three) thermistors properly combined in a single package (Fig. 4-7) provide good linearization over a wide temperature range (up to 100°C). Each thermistor effectively covers part of the range. The matching of the close-tolerance thermistors and resistors is an involved and somewhat costly process normally done by the manufacturer.

4-3   Bridge Amplifier
for Resistance Sensors

The bridge amplifier (Fig. 2-16) discussed in Chap. 2 is well suited as a readout device for a resistance thermometer or any resistance-type sensor. Bridge voltage is derived from the precision voltage regulator

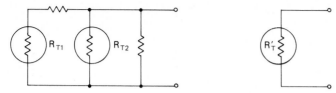

Figure 4-7 Two-thermistor linearization: (*a*) actual circuit and (*b*) equivalent circuit.

through an amplifier. The bridge voltage and resistance are chosen such that the sensor power dissipation is not excessive.

From the general expression for the bridge sensitivity [Eq. (2-15)] and the thermistor temperature coefficient [Eq. (4-2)], the circuit response for the bridge near balance (ΔT small) is

$$V_o = S_T \, \Delta T \tag{4-7}$$

where $S_T = AV_b/4$ and A = the amplifier voltage gain.

The bridge readout (Fig. 4-8) can be operated in either of two modes, a null mode or a small-deviation (limited temperature excursion) mode. The bridge output (ΔV), which is balanced with respect to ground, is converted to a single-ended output (v_o) by an instrumentation amplifier. With the null mode, R_x is adjusted until the output voltage v_o is zero. The value of R_T is read off the dial, an inconvenience, and the temperature found by referring to a calibration table. With the second mode of operation, R_x becomes null at the lowest temperature of a limited range (e.g., 95°F for 95 to 105°F). The amplification factor is chosen so that the output voltage change is an even number times the temperature change (e.g., 0.10 V/°F). An advantage of this circuit is that it can measure small changes in temperature accurately, but it has the disadvantage of a limited linear range.

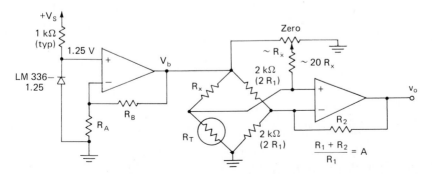

Figure 4-8 Bridge amplifier for resistance sensors.

When $\delta = \Delta R/R_x$ is not small, the output v_o contains quadratic and cubic terms also. From Eq. (2-14) the response becomes

$$v_o \approx \frac{AV_b}{4}\left[\delta - \frac{\delta^2}{2} + \frac{\delta^3}{4}\right] \qquad (4\text{-}8)$$

4-4   Resistance Thermal Detector (RTD)

The RTD is made of a metal with a positive temperature coefficient (α). The resistance variation is an order of magnitude less for that of the thermistor, but the response is much closer to linear. Platinum is usually used because of its wide temperature range, stability, and resistance to chemical attack. Sometimes, however, copper and nickel are used (Fig. 4-9).

Both wire-wound and thin-film versions of the platinum RTD are available. Wires are wound strain-free around an insulating form and protected in a housing such as a stainless steel tube (Fig. 4-10b). Two (or more, usually copper) lead wires are brought out of the probe. The resistance (R_T) of an RTD as a function of temperature (T), in degrees Celsius, is approximated by the linear equation

$$R_T = R_o\,(1 + \alpha T) \qquad (4\text{-}9)$$

where R_o is the RTD resistance at 0°C and α is the temperature

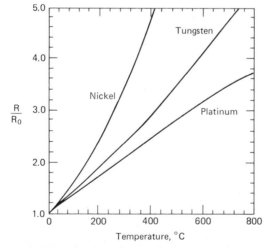

Figure 4-9 Resistance versus temperature for various RTD metals.

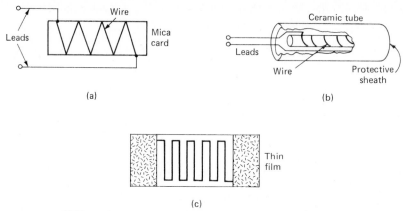

Figure 4-10 RTD configurations: (a) spiral wire wound in probe; (b) filar wound; and (c) thin-film.

coefficient. Note that α is defined differently here than for the thermistor, since ΔT here is large and R_o is fixed (not R_T). For pure platinum, the value of α (averaged over 0 to 100°C) is 0.003927 $\Omega/\Omega \cdot$ °C. More commonly, the RTD is made of platinum with impurities added so that $\alpha = 0.00385$ (0 to 100°C average), an established standard. A more accurate equation for the resistance variation with temperature is given by

$$R_T = R_o (1 + \Delta T - BT^2) \tag{4-10}$$

where $A = 3.986 \times 10^{-3}$ and $B = 5.88 \times 10^{-7}$ for $\alpha = 0.003927$, or $A = 3.90 \times 10^3$ and $B = 5.7 \times 10^{-7}$ for $\alpha = 0.00385$ with T in degrees Celsius and R_T a best-fit (0.1 to 0.5°C) approximation in the range 0 to 420°C. The resistance-to-temperature conversion* is given in tabular form in Table 4-2 and in graphical form in Fig. 4-9.

A close tolerance of 0.2 percent for resistance is standard (class B), thus facilitating interchangeability between sensors. Long-term stability of wire-type platinum RTDs is exceptional (0.01°C precision possible).

Thin-film platinum RTDs (Fig. 4-10c) are smaller, lower cost, and higher in resistance ($R_o = 1000\ \Omega$) than the wire type. High-resistance units have less internal heating and less lead resistance error. Frequently they are manufactured with the standard temperature coefficient ($\alpha = 0.00385$). Stability over long periods and with multiple

---

*International Practical Temperatures Scale established by the International Electrochemical Commission (1968). For a discussion, see Robert P. Benedict, "Fundamentals of Temperature, Pressure, and Flow Measurements," 2d ed. John Wiley, New York, 1977.

TABLE 4-2   Resistance-to-Temperature Conversion Chart for a Platinum RTD (α = 0.00385, R_o = 100 Ω)

T, °C	R_T, Ω	T, °C	R_T, Ω
−250	27.08	+500	280.90
−100	60.25	+600	313.59
±0	100.00	+700	345.13
+100	138.5	+800	375.51
+200	175.84	+900	(404.5)
+300	212.02	+1000	(432.2)

temperature cycling is very good but still inferior to the wire version, especially at high temperatures.

Suitable readouts for the RTD are the Wheatstone bridge with instrumentation amplifier (Fig. 2-18), the resistance-to-voltage converter (Fig. 4-5) and the bridge amplifier (Fig. 4-8). A low offset voltage (V_{os}) and low drift at V_{os} with temperature are important considerations for the op-amp choice. Bridge circuits are best for small resistance changes (bridge near balance) because they are less sensitive to bridge voltage (output proportional to ΔR rather than R). However, the bridge is nonlinear for large sensor resistance changes, and thus for most applications the R-to-V converter is preferable. Sensor voltage and current must be limited (2 V max for R_o = 100 Ω) to avoid sensor internal heating and op-amp output current saturation.

Lead resistances (R_W) can result in an appreciable error (R_W = 2.4 Ω for a 50-m, 22-gage copper wire), especially for low resistance (R_o = 100 Ω) RTDs. The three-wire connection with associated bridge (Fig. 4-11) provides compensation, assuming the lead resistances are small and equal. The cancellation of the small lead voltage drops at the amplifier input may be understood by noting that at balance (R_T = R_B), $v_1 = v_T + 3v_W = v_2$, where v_W is voltage drop on the nongrounded lead. Excitation at the bridge by a constant current (I_B) rather than a constant voltage (V_B) source voltage improves accuracy since the effective bridge voltage is reduced by the ratio $3R_W/2R_o$ for constant-voltage drive. Lead resistance errors are small if R_o is large, and therefore the three-wire technique is usually not needed for thin-film-type RTDs.

Although the platinum RTD is rather linear (only ±0.5°C nonlinear in 0 to 100°C range), the remaining nonlinearity, a slightly downward curvature, can be corrected by increasing the bridge voltage as the temperature increases (Fig. 4-12). This small positive feedback also can compensate for bridge nonlinearity and, more generally, for all second-order nonlinearities (see Sec. 4-3).

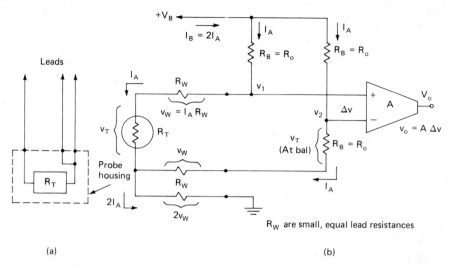

(a)   (b)

Figure 4-11 Three-wire RTD readout: (*a*) lead connection to sensor and (*b*) bridge.

4-5 Thermocouples

A thermocouple is made from two dissimilar conductors, usually metal wires. When the junction is heated, a small thermoelectric voltage is produced which increases approximately linearly with junction temperature. Actually two junctions are always present, a measurement junction and a reference junction, as indicated in Fig. 4-13. At any junction between two metals a potential difference v_{12} exists as a consequence of differences in the effective concentration of electrons in

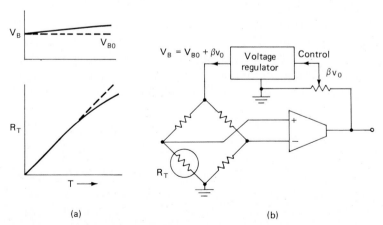

(a)   (b)

Figure 4-12 Nonlinearity compensation: (*a*) RTD response with nonlinearity exaggerated; (*b*) block diagram of correction circuit.

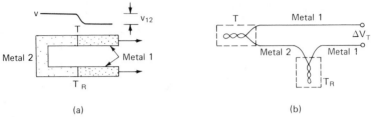

(a)   (b)

Figure 4-13 Thermocouple junctions: (*a*) diagram showing point of potential difference; and (*b*) measurement and reference junction temperature regions.

the two metals. The effect is similar to the one that produces the internal bias voltage across diode junctions but is smaller. The two junctions of Fig. 4-13*a* are connected into the circuit with opposite polarity. When they are the same temperature, the two junction potentials cancel ($\Delta V_T = v_{12} - v_{21} = 0$), but otherwise ΔV_T may be positive or negative, depending on whether T is higher or lower than T_R. In physics laboratories an ice bath (0°C) is normally used as a reference, and thermocouple tables are referenced to 0°C measuring the voltage.

Often a third metal is introduced into the thermocouple circuit in practical applications. If nothing else, the effect of the copper leads which connect the thermocouple to the voltmeter or amplifier should be considered. Fortunately, a law on the summation of thermocouple potentials, derivable from general thermodynamic relations, exists. In equation form, refering to the circuit of Fig. 4-14*a*, it states:

$$\Delta V = V_{AB}(T) + V_{CA}(T_R) + V_{BC}(T_R)$$
$$= V_{AB}(T) - V_{BA}(T_R) \tag{4-11}$$

Thus if the two reference junctions of the thermocouple wires (A and B) are made to a third type of wire (C, usually copper) *at the same temperature* (T_R), then the voltage between the two lead wires (C) is exactly the same as if the junction were between metals A and B alone. If the two thermocouple wires are connected directly to the input

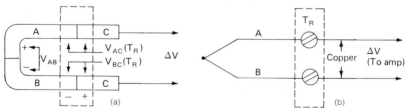

Figure 4-14 Reference junction voltage relations: (*a*) voltage summation and cancellation for three metals; (*b*) reference junction block.

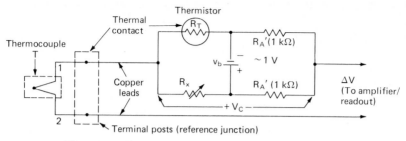

Figure 4-15 Thermocouple junction compensation bridge.

terminals, both of which are at the reference temperature T_R, the voltage of these terminals (ΔV) will be (Fig. 4-14b)

$$\Delta V = V_T^\circ - V_J^\circ \tag{4-12}$$

where V_T° is the measurement thermocouple table voltage at temperature T and V_J° is the reference junction voltage at temperature T_R. Both V_T° and V_J° are taken from a standard thermocouple table with a 0°C reference (actually ΔV is independent of the table reference temperature).

Often the junction temperature, around room temperature, is measured with another thermosensor (thermistor or IC type). The corresponding junction voltage (V_J°) is found by looking it up in the thermocouple table (backward), perhaps by microprocessor. Then V_J° is added to the measured voltage (ΔV) to obtain V_T°, the thermocouple voltage as if the reference junction is at 0°C [Eq. (4-12)]. The temperature T is then found from the table (0°C reference).

In commercial instruments a thermistor monitors the terminal temperature and provides a temperature-dependent voltage which just compensates for the junction voltage (Fig. 4-15). It is necessary to choose the thermistor temperature coefficient and/or battery voltage V_b such that V_J matches the thermocouple temperature coefficient in millivolts per degree Celsius. The zeroing resistor R_x is set so that the output voltage Δv is zero when the thermocouple is at 0°C.

Table 4-3 lists the characteristics of several standard thermocouple types. The Instrument Society of America (ISA) letter designations (e.g., K, J, R) are indicated. Thermocouple types differ in temperature range, resistance to chemical attack, stability, and accuracy. Conversion from voltage to temperature is done by means of tables (Table 4-4), polynomial equations, or graphs (Fig. 4-16).

The CAZ amplifier (Fig. 4-17) is an excellent preamplifier because of its low offset voltage ($<5~\mu$V). Instrumentation amplifiers, which have both inputs ungrounded, are often used since the measurement thermocouple may be grounded for better thermal contact.

TABLE 4-3   Thermocouple Characteristics for Several Metal Couples

Thermocouple type	ISA type	Temperature range, °C	Approximate tolerance*	Comments
Chromel/Alumel	K	−270–1370	±2.2°C/0.75%	Fairly inert, good range
Chromel/Constantan	E	−270–1000	±1.7°C/0.5%	High output, but can oxidize
Copper/Constantan	T	−270–400	±1°C/0.75%	Accurate, fairly inert, but short range
Iron/Constantan	J	−200–760	±2.2°C/0.75%	Stable but oxidizes over 600°C
Platinum 13% Rhodium/platinum	R	0–1760	±1.5°C/0.25%	Stable, inert, but low output, high temperature

* Error limit of standard grade; for premium grade, divide by 2, for extension grade, multiply by 2.

4-6   Diodes as Thermosensors

Ordinary silicon diodes may be used as temperature sensors in undemanding applications by taking advantage of the almost linear decrease in diode voltage (V_D) with temperature (Fig. 4-18a). A linear voltage output is produced by the circuit of Fig. 4-18b. Switching at a fixed temperature may be done using a comparator (Fig. 5-34).

4-7   IC Thermosensor

The IC thermosensor (Fig. 4-19) is based on the diode temperature dependence. One version (AD590) has a current output (I_T) proportional to absolute temperature. With another (LM335) the output voltage (V_T) is proportional to temperature. Its linearity is excellent (±0.1 percent) over the full range (−40 to +175°C for metal case), but since the output (current or voltage) is proportional to temperature in kelvin for most types currently available, a large zero offset is necessary to obtain a readout in degrees Celsius or Fahrenheit. The

TABLE 4-4   Condensed Tables for Two Standard Thermocouples

Temp, °C	Type E, mV	Type R, mV
−100	−5.24	—
±0	0.00	0.00
+100	6.32	0.65
200	13.42	1.47
300	21.03	2.40
400	28.94	3.41
500	37.00	4.47
600	45.08	5.58

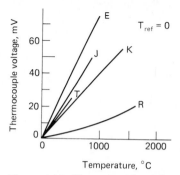

Figure 4-16 Thermocouple voltage versus temperature for several common metal couples.

scale factor for the voltage output is set by the potentiometer (Fig. 4-19*b*), typically to 0.10 V/°C. Newer IC thermosensors (e.g., LM34 or LM35) have an internal compensation.

Connection of the sensor to a positive supply is all that is needed for the current-type sensor. While a load resistor may be used to convert the sensor current to voltage, the current-to-voltage converter (Fig. 4-20) is better because it provides the necessary stable zero offset. Current-type sensors are less sensitive to noise pickup and lead resistance than the voltage-output type (Fig. 4-19*b*), an advantage if the sensor is located remote from the readout.

4-8 Less Common Thermosensors

Piezoelectric crystals resonate at a frequency which is temperature-dependent. When they are made a part of an oscillator circuit (Chap. 11), the shift in frequency is approximately proportional to tempera-

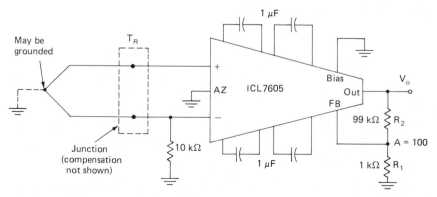

Figure 4-17 Thermocouple voltage CAZ amplifier.

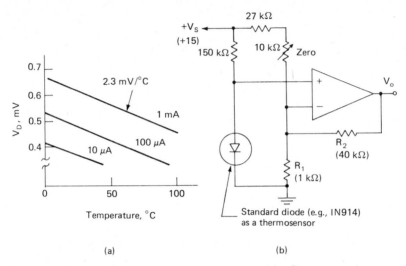

Figure 4-18 Silicon diode as a temperature sensor: (a) voltage versus temperature and (b) linear amplifier.

ture. Often quartz crystals specially cut to maximize the temperature
dependence are chosen because the frequency shift with temperature is
quite linear and reproducible (the temperature dependence is minimized in conventional cuts). A further advantage is the direct digital
readout possible. When the circuit is properly designed, the frequency
shift (or at least the number of cycles counted) is numerically equal to
the temperature in degrees Celsius. The shift with the frequency is
rather small (1 kHz/°C at 10 MHz), and therefore a costly stable
reference oscillator and frequency counter must be employed. Aside

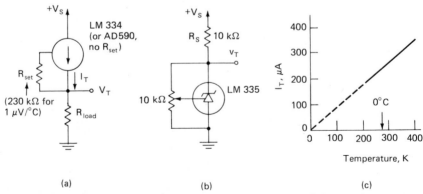

Figure 4-19 IC thermosensor: (a) basic current-output circuit; (b) basic voltage-output
circuit; and (c) response of linear version.

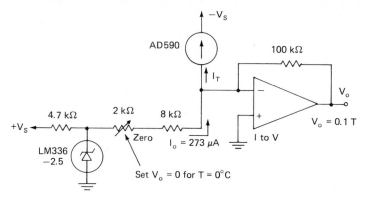

Figure 4-20 Current-to-voltage converter readout for IC thermosensor.

from cost, the crystal sensor has the disadvantage of a relatively large size and long response time.

Temperature-dependent capacitors are sometimes employed as thermosensors. Like the piezoelectric crystals, the capacitor is made a part of an oscillator-resonance circuit, so that the frequency shift is a function of temperature. A Colpitts oscillator and ceramic capacitor work well for applications where simplicity is desired but high accuracy is not required. For example, a miniaturized version has been strapped to a bird so that the oscillator frequency, monitored by a radio receiver, indicated its temperature in flight. Available capacitors seem either to have a very small temperature coefficient or are not highly reproducible and therefore are not satisfactory as thermosensors for exacting applications.

Chapter

5

Electrooptical Devices

5-1 Optical Spectra and Energy Relations

The choice of optical detectors and sources is strongly influenced by the wavelength of light required. Many detectors, for example, will not respond at all if the wavelength is longer than a certain cutoff wavelength. Except for the insensitive bolometer, all optical detectors involve a quantum effect, in which the photon energy is an important parameter. The energy of the photon E_p is given by

$$E_p = \frac{ch}{\lambda} \tag{5-1}$$

where h = Planck's constant = 6.62×10^{-34} J · s, λ = wavelength of light, and c = velocity of light = 3.0×10^8 m/s. It is convenient to express photon energy in electron-volts (eV) and wavelength in nanometers (nm); in this case, the constant ch is equal to 1240 eV/nm. Nearly all electrooptical effects require that a single photon have sufficient energy to free an electron by overcoming the energy of a chemical bond or trap in a solid. Since most bond energies lie in the range of 0.2 to 4 eV, the minimum photon energy is in this range. Actually trapping energies lower than this are common in semiconductors, but the trapped charge carriers are released by the thermal energy (0.025 eV at room temperature) unless the detector is cooled, and therefore light has little additional effect. In Fig. 5-1 the range of wavelengths, or optical spectrum, is indicated.

Light intensity referring to the amount of power emitted from the source is commonly expressed in watts. Intensity can also be defined as an energy or radiant flux, expressed in watts per square meter. If, as is

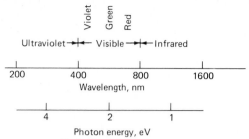

Figure 5-1   Optical spectrum.

usually the case, the intensity varies with wavelength, the power per unit wavelength, or spectral power density, is often specified as a function of wavelength, as in Fig. 5-2.

Sometimes light intensity is expressed in photometric units, such as candelas (cd) or lumens (lm), which are based on effective brightness as perceived by the eye. Because the sensitivity of the eye depends on the wavelength, no constant ratio exists between the photometric units and fundamental units such as power. At 550 nm where the eye is most sensitive (Fig. 5-2) the ratio (luminous efficiency) is approximately 680 lm/W. A one-candela point source results in a flux of one lumen through a one-square-foot area on a sphere with a radius of one foot. Another way of measuring light intensity is to specify the numbers of photons of incident power emitted per unit time. As indicated by Eq. (5-1), the number of photons per watt depends on the wavelength. At 555 nm, the light beam of 1 W corresponds to 2.8×10^{34} photons per second. Definitions and interconversion of a number of units are indicated in Table 5-1.

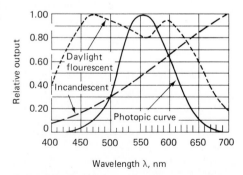

Figure 5-2 Spectral power density: relative response of the eye (phototropic response) when compared with common room-light sources.

TABLE 5-1 Definitions of Optical Units*

Parameter	Symbol	Definitions	Units
Radiometric			
Radiant energy	Q_e		erg, J, cal, kWh
Radiant flux	P	$P = dQ_e/dt$	erg/s, W
Radiant emittance[†]	W	$W = dP/dA$	W/cm², W/m²
Irradiance[†]	H	$H = dP/dA$	W/cm², W/m²
Radiant intensity	J	$J = dP/d\omega$ where ω = solid angle through which flux from point source is radiated	W/sr
Radiance	N	$N = d^2P/d\omega\,(dA\cos\theta)$ $= dJ/d4\cos\theta$ where θ = angle between line of sight and normal to surface considered	W/sr · cm² W/sr · m²
Photometric			
Luminous efficacy	K	$K = F/W$	lm/W
Luminous efficiency	V	$V = K/K_{\max}$	
Luminous energy (quantity of light)	Q_r	$Q_r = \int_{380}^{760} K(\lambda)Q_e\lambda\,d\lambda$	lm ·h , lm · s (= talbot)
Luminous flux	F	$F = dQ_r/dt$	lm
Radiometric			
Luminous emittance[†]	L	$L = dF/dA$	lm/ft²
Illumination (illuminance)[†]	E	$E = dF/dA$	footcandle (fc = lm/ft²), lux (1x = lm/m²), phot (ph = lm/cm²)
Luminous intensity (candlepower)	I	$I = dF/d\omega$	candela (cd = lm/sr)
Luminance (brightness)	B	$B = d^2F/d\omega(dA\cos\theta)$ $= dI/(dA\cos\theta)$	cd/in², etc., stilb (sb = cd/cm²), footlambert [ft · L = cd/(π · ft²)], apostilb [asb = cd/(π · m²)]. L [= cd/(π · cm²)]

Source: Table from the "American National Standard Nomenclature and Definitions for Illuminating Engineering."
* Used by permission of the Illuminating Engineering Society.
[†] W and L refer to "emitted from," while H and E refer to "incident on."

Most photodetectors are specified in terms of irradiance with units of watts per square centimeter.

5-2   Photodiodes

The principle upon which the photodiode operates is a quantum effect. Consider the reverse-biased *pn* junction sketched in Fig. 5-3a operated as a photodiode. If light is absorbed in a semiconductor, with a certain probability an electron will be excited into the conduction band. The extra charge carriers, if produced far from the junction, cause a slightly increased conductivity or photoconductivity, which is negligible if the diode is reverse-biased. Minority carriers produced near the junction, in the depletion region, however, have a marked effect because they are drawn by the field through the junction and contribute to the reverse current, which is very small in the dark. The net effect is that the reverse current i_p of the photodiode is proportional to the light intensity W; that is,

$$i_p = K_s W \tag{5-2}$$

where K_s is the sensitivity factor (Fig. 5-4a). The junction must be close to the surface (*p* region thin in Fig. 5-3b), since light is absorbed as it passes through silicon (energy absorption is the first step in a photoelectric conversion). Since standard photodiode chips are small, a glass or plastic lens is often used to focus the light onto the active area (Fig. 5-4b).

Typical photodiode characteristics are shown in Fig. 5-5. Note that the current is linearly proportional to the incident light intensity over a wide range. It must be emphasized that it is the reverse current, and not the conductivity, of the device that is proportional to light intensity. The current at a given light level is nearly independent of diode

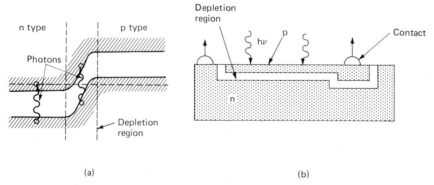

(a)   (b)

Figure 5-3   Light interaction in a semiconductor: (*a*) energy diagram; (*b*) photodiode geometry.

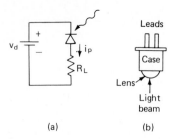

Figure 5-4 Basic photodiodes:
(*a*) diode bias circuit and (*b*)
case with lens.

voltage, and any bias voltage higher than that about 0.5 V will do.
Dark currents are a small fraction of total current (typically 0.01 μA
for a full scale of 500 μA).

Several photodiode (and phototransistor) responses are shown in
Fig. 5-5*b*. Silicon detector response cuts off at wavelengths longer than
1100 nm because of the band-gap energy limitation. Standard detectors peak at about 800 to 900 nm (red to near-infrared) but maintain

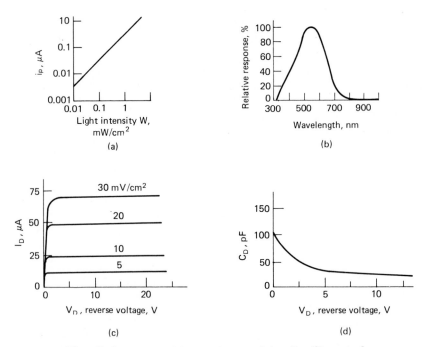

Figure 5-5 Photodiode response: (*a*) current versus intensity; (*b*) spectral response;
(*c*) current versus voltage; and (*d*) capacitance versus voltage.

good (but not constant) sensitivity through the full visible region. Blue-enhanced detectors have a flatter wavelength response. The short wavelength cutoff in the ultraviolet (UV) is due in part to the lens or window, but the rapidly increasing light absorption below 350 to 400 nm limits the response in any case to near-UV. Gallium arsenside photodiodes, developed for fiber-optic applications, respond further into the infrared (peak at 1500 nm). Germanium and germanium-doped and a number of other low-energy-gap photodiodes respond even further into the infrared (2000 to 20,000 nm) but may require cooling to low temperatures (perhaps 77 K) in order to reduce the thermally produced dark current.

As indicated in Fig. 5-5c, photodiode (dc) current is practically independent of reverse voltage (but depends on light intensity). Thus it acts as a constant current source. Diode capacitance (C_D), however, decreases substantially with voltage because the width of the depletion region increases with voltage. If a high value load resistor is used (Fig. 5-6a), the response time can be adversely effected.

The current-to-voltage converter (Fig. 5-6a) is an ideal amplifier for the photodiode current. A simple load resistor R_L (Fig. 5-6a) will work, but since it must have a high value (100 kΩ to 100 MΩ) to achieve high sensitivity, a high-input-resistance (FET) amplifier is then needed as an interface. Note that the response time, $\tau_r = R_L C_D$, is limited when R_L is high. The current-to-voltage amplifier has the same gain and number of components but has a faster response since the stage input impedance is near zero (because $v_- \approx 0$ for infinite-gain approximation). Sometimes a zero or dark current offset control is added, but the variability of the dark current with temperature and time limits its usefulness.

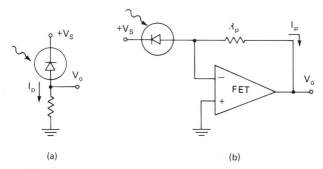

(a)   (b)

Figure 5-6 Photodiode readout: (a) load resistor; (b) current-to-voltage converter.

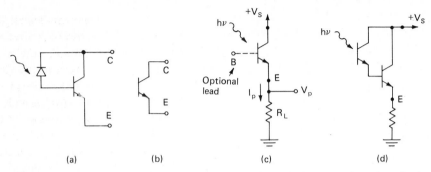

Figure 5-7   Phototransistors: (*a*) equivalent circuit; (*b*) symbol; (*c*) simple resistor load; and (*d*) Darlington type.

5-3   Phototransistors

A phototransistor is a combination of a photodiode and phototransistor on the same chip (Fig. 5-7*a*). Diode current is amplified by the transistor section so that the phototransistor output current ($I_p = \beta I_D$) is typically 100 times larger than the photodiode current. The higher gain is accompanied by nonlinearity, a consequence of the variation of β with I_E. Also, the response time is less. Thus the phototransistor is not as good as the photodiode where linearity and speed are important. The higher phototransistor current gain is of no advantage with the readout of Fig. 5-6*b*, since the stage can be made as high as desired by increasing R_p.

The phototransistor works well for interrupted light beam applications. A load resistor followed by a comparator or Schmitt trigger converts the switched signal to digital logic level. Relatively high output currents, as well as high sensitivity, are provided by phototransistors, especially the Darlington type (Fig. 5-7*d*). A relay might be driven directly. Other transistor output configurations are discussed in connection with optical isolators and fiber optic detectors (Sec. 5-10).

5-4   Photovoltaic Cells

Photovoltaic cells produce a voltage and/or current when illuminated by light. The principle is the same as the photodiode and they have the same construction (Fig. 5-3), except that the photovoltaic cell usually has a much larger area. The same device, in fact, may be operated in the photodiode or the photocell mode. In contrast to solar cells used for power production, the photovoltaic cells used for light-intensity measurement are operated short-circuited (Fig. 5-8*a*) because the short-circuit current (I_{sc}) is then linearly proportional to light intensity. Reverse and short-circuit currents are nearly identical for the same

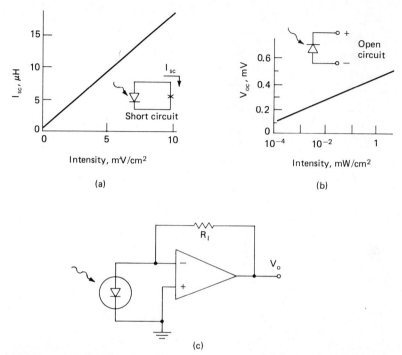

Figure 5-8   Photovoltaic cell configuration: (*a*) short-circuit current versus light intensity; (*b*) open-circuit voltage versus intensity; and (*c*) current-to-voltage converter readout.

device and conditions. The current-to-voltage converter (Fig. 5-8*c*) provides the necessary zero input impedance.

Solar cells are large-area photovoltaic cells intended for power generation. Cells are available which will deliver a load current of 1 mA/cm with a terminal voltage of 0.6 V when exposed to average sunlight.

5-5   Light-Emitting Diodes

The principle of the light-emitting diode (LED) is the inverse of the photodiode. When the junction is forward-biased, charge carriers flow across to recombine with the majority carriers and release energy in the form of photons with a photon energy comparable to that of the band-gap energy. Semiconductors with higher gap energies result in higher photon energies, i.e., light of shorter wavelengths. For silicon diodes, the wavelength maximum is about 900 nm, which lies in the near-infrared. For semiconductors with higher band-gap energies, such as GaAs, the emission is in the visible region. Currently available visible LEDs are red, orange, yellow, green, and (rarely) blue. Emission spectra are indicated in Fig. 5-9*b*.

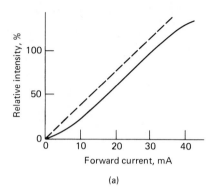

(a)

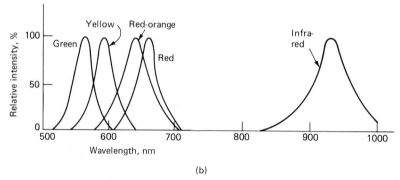

(b)

Figure 5-9 LED characteristics: (*a*) emitted spectrum; (*b*) emitted intensity versus current.

The light output increases approximately linearly with forward current, but some nonlinearity is noticeable at low and high current levels (Fig. 5-9*a*). It can be operated in pulsed mode such that the average power does not exceed the maximum power dissipation of the device. The permissible peak current is usually 5 to 20 times the average or direct current, which is typically 10 to 50 mA. The efficiency drops with increasing temperature, including the temperature rise due to internal heating.

Current-voltage characteristics of LEDs (Fig. 5-10) are similar to those of any forward-biased diode except that appreciable current does not flow through visible LEDs until 1.4 to 2.7 V is applied (both conduction threshold voltage and photon energy increases with band-gap energy). In order to limit the current flow to a safe value, a series resistor (R_S) is ordinarily required. Connection to a fixed voltage source may destroy the diode. The reverse voltage is also rather small, typically 3 to 10 V.

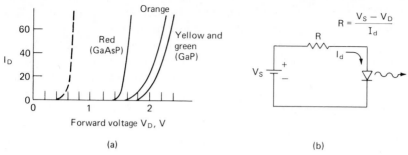

(a)

(b)

Figure 5-10 LED electrical characteristics: (a) current versus voltage; (b) test circuit.

5-6 Semiconductor Photocells

A semiconductor photocell, an older device, is based on photoconductivity. In contrast to the photodiode, which acts as a current source, the photocell conductivity (G_p) increases with light intensity (Fig. 5-11b). The resistance of a cell may change from over 10 MΩ in the dark to

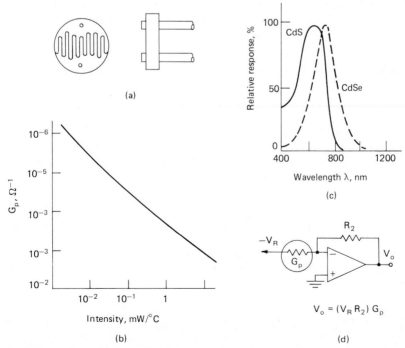

(a)

(b)

(c)

(d)

$$V_o = (V_R R_2) G_p$$

Figure 5-11 Semiconductor photocell: (a) geometry; (b) intensity response; (c) spectral response; and (d) conductance-to-voltage readout.

under 10 Ω in a bright light. Response time is much longer than the photodiode, often requiring over 10 ms to return to the dark conductance level. Devices with large sensitive areas are less costly than the corresponding size photovoltaic cell. Photoconductors have poorer sensitivity, stability, and speed than photodiodes. They are used mostly where cost is a factor.

Conversion to voltage is best done with the circuit of Fig. 5-11d. A simple load resistor similar to that of Fig. 5-6a is an alternative. Dark conductance can be subtracted with a zero offset circuit but the temperature sensitivity limits its value.

5-7   Photomultipliers

A photomultiplier is based on the photoelectric effect, i.e., the release of electrons from a metal surface hit by higher-energy photons. Its operating principle is basically similar to that of the nearly obsolete vacuum tube photocell (Fig. 5-12). When a photon strikes a surface, an electron is released if its energy ($E = h\nu$) exceeds that of the work function of the cathode surface. The electrons are collected by a positively charged anode ($v_A \approx 150$ V) resulting in a photocurrent i_p proportional to the number of photons, i.e., to the intensity of light at any particular wavelength. Because the current is small, a sensitive current amplifier is required unless light is very intense. Another disadvantage is that lower-energy photons, specifically at the red and infrared end of the spectrum, cannot be detected because the work function of most available materials is comparatively high, over 1 eV. An infrared photomultiplier ($\lambda < 1100$ nm) is, however, available. Miniature vacuum tube photocells, although insensitive, are very fast and suitable for monitoring the intensity of laser pulses.

Adding a multistage electron multiplier to a photocell produces a photomultiplier (Fig. 5-13). Photons of sufficient energy striking the cathode release electrons, which are attracted to the neighboring dynode because of its positive potential with respect to cathode. The potential difference is fairly high (50 to 200 V), accelerating the electrons so that as they strike the dynode, secondary electrons are emitted. Typically two or three low-energy secondary electrons are

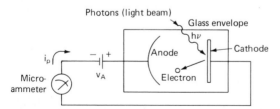

Figure 5-12   Vacuum tube photocell schematic.

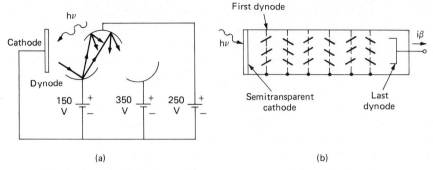

Figure 5-13 Photomultipliers: (*a*) schematic of first stages; (*b*) end-on multiplier.

emitted for each incident electron, thus producing an electron-multiplying effect. This occurs at each stage or dynode. The result is that the photocurrent emitted by the cathode is multiplied by $A_m = (A_s)^n$, where A_s is the stage current gain (ratio of secondary to incident electrons) and n is the number of stages, typically 7 to 14. Gains A_m of 10^5 or 10^6 are common and account for the high sensitivity of the photomultipliers.

An end-on photomultiplier (Fig. 5-13*b*) has a semitransparent photocathode film on the end window. Light passing through the glass causes an electron to be released from the backside of the film. It is drawn by focusing electrodes (not shown) to the venetian-blind-type dynodes, where secondary electron emission occurs as before.

Voltage is applied to the dynodes and cathode from a negative high-voltage source V_H through a resistor chain (Fig. 5-14*a*). A positive ground is required because the load resistance R_L and/or amplifier common must be connected to the power-supply positive terminal. The secondary electron emission is voltage-dependent, and the total gain A_m depends on the nth power of stage gain, so that the total gain is strongly dependent on the applied voltage V_H. As in the photodiode, the photomultiplier output is a current source, and thus

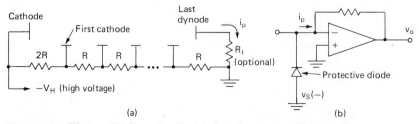

Figure 5-14 Photomultiplier circuit: (*a*) dynode voltage divider; (*b*) current-to-voltage converter for mid to high light levels.

the input amplifiers described for the photodiode are suitable here. Diodes to protect against high-voltage surges (due to exposure to high light intensities) are suggested.

The photomultiplier is the fastest and most sensitive photodetector available. It will respond to single photons with high efficiency. At low light levels, where the single photon pulses do not overlap, pulse-counting techniques can be used to measure light intensity. If the load resistance R_L is low (50 to 1000 Ω), the pulse width is typically 5 to 50 ns ($\tau_{rise} < 1$ ms). The photomultiplier will respond to UV ($\lambda < 200$ to 400 nm) if the envelope is UV transparent (quartz). The photomultiplier cathode can be damaged by exposure to higher light intensities when the high voltage is on. The fragility, size, and cost of the tube and need for a regulated high-voltage supply are substantial disadvantages.

5-8   Photodiodes and LED Lenses

Photodiodes and phototransistors often employ lenses to focus an incoming divergence or broad light beam onto the small sensitive area (Fig. 5-15a). Similarly, the light from the small emitting area of an LED may be focused by a built-in lens into a beam of greater or lesser divergence. The location of an image (S_2) is given by the lens formula, following the definitions in Fig. 5-15b:

$$\frac{1}{S_2} = \frac{1}{f} - \frac{1}{S_1} \tag{5-3}$$

If S_1 is negative, the beam will diverge (virtual image). A parallel or collimated beam corresponds to $S_2 = \infty$, which occurs if the source S_1 is placed at the focal point f of the lens. Since the LED emitting area is not a point, although small, the beam will diverge slightly even with the best placement. Small plastic lenses also produce appreciable aberration.

Photodetectors with lenses are best mounted in a recessed holder (Fig. 5-16a) which will block out stray light. It is important to reduce the effects of general room lighting (120 Hz for fluorescent lights on a 60-Hz power line) or perhaps sunlight (dc). Optical (color) filters which have a peak transmission at the wavelength of the incoming beam will further enhance the signal-to-noise ratio. In most optical systems, sensitivity is less of a problem than background interference because the amplifier gain can be made as high as desired. Modulation of the optical beam at a high frequency (10 to 200 kHz), together with tuned amplifiers or demodulators, are effective in eliminating the adverse effects of background illumination.

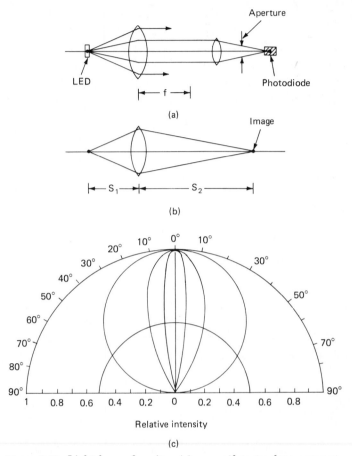

Figure 5-15 Light beam focusing: (*a*) source/detector lens arrangement; (*b*) simple lens imaging; and (*c*) radiant intensity or sensitivity distribution curve.

5-9 Optical Pyrometers and Thermal Radiation Detectors

The bolometer is a broadband sensor for electromagnetic radiation (Fig. 5-17*a*). Bolometers are identical to semiconductor photocells in many respects, but the mechanism responsible for the conductivity change depends simply on the temperature rise due to absorption of radiation. A thermistor or thermopile (10 to 50 small thermocouples in series) may be used to measure the temperature rise. Electromagnetic energy absorbed at any wavelength, including microwave or radiofrequency, will have the same effect. This broad frequency re-

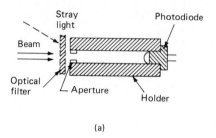

(a)

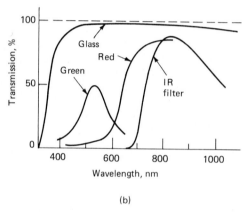

(b)

Figure 5-16 Photodetector mounting: (*a*) recessed holder; (*b*) optical filter transmission characteristics.

sponse is not approached in other optical detectors and is its most important advantage. Sensitivity and response time are rather poor, but accurate calibration can be made by comparing the response to a standard amount of electrical power in an attached resistor. If a lens

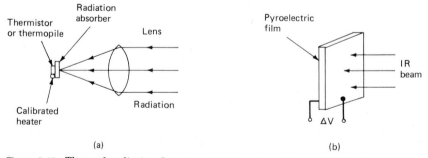

(a)   (b)

Figure 5-17 Thermal radiation detectors: (*a*) bolometer; (*b*) pyroelectric film.

transparent at the wavelength of interest cannot be obtained, a focusing mirror may be substituted.

A related detector is a pyroelectric film (e.g., polyvinylidine fluoride) which generates a voltage upon radiation (Fig. 5-17b). The pyroelectric response is associated with piezoelectric materials which, along with readouts, are discussed in Chap. 6. The response occurs when a light beam, usually infrared, strikes the surface. Like all piezoelectric devices, the detector responds only to changes in signal level and not to dc. Sensitivity variation is a problem.

5-10 Optical Isolators and Fiber-Optic Transmitters and Receivers

An optical isolator or coupler consist of an LED and photodiode mounted in the same package. The photodiode is mounted to capture the maximum amount of light from the LED; that is, the two diodes are optically coupled but electrically isolated. These devices are useful in isolating grounds in larger systems and where high-voltage isolation between circuits is required. Some isolators have a built-in transistor, which can be operated in either the common-collector mode or the common-emitter mode. Typical characteristics are shown in Fig. 5-18. Because of the slightly nonlinear LED characteristics, the current gain ($\beta = i_o/i_i$) varies with the current. Current gains of 0.1 to 10, frequency response beyond 1 MHz, and insulation of 1 to 3 kV are typical.

A voltage-to-current amplifier (Chap. 2) is a convenient driver for the LED isolators (Fig. 5-19), and the current-to-voltage converter is a convenient receiver. Only positive signals are allowed. Resistor R_b and diode D1 limit the driver amplifier voltage and current swing. Because

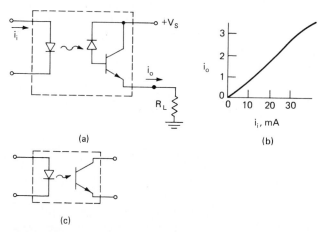

(a)

(b)

(c)

Figure 5-18 Optical isolator: (a) standard phototransistor connection and (b) transfer characteristics.

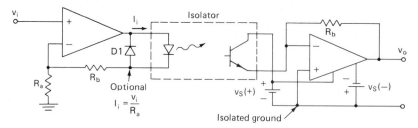

Figure 5-19 Optical isolator analog driver and receiver.

no feedback exists here between the receiver and driver, the overall voltage gain will vary somewhat with temperature and with the individual isolator current gain. Of course, the power supplies for the driver and receiver sections must be separated if ground isolation is required. If batteries are impractical, an isolated dc-to-dc power supply (Chap. 21) can be used.

Optical couplers or isolators which combine an incandescent (or neon) bulb and photocell in one package are also available. They are similar in function to the LED photodiode devices discussed above, but their response time is much slower. While in most applications the LED type is superior, for others, such as automatic-gain-control circuits, the linear current-voltage characteristic of the photocell (a resistance, not a diode) is important.

The isolator with silicon-controlled-rectifier (SCR) and triac drivers, used for power control, are discussed in Chap. 22.

Fiber-optic links (Fig. 5-20) are similar to optical isolators except that the LED transmitter and photodiode receiver are separated by a transparent glass, or sometimes plastic, fiber. Light entering at a small angle will be internally reflected at the core-cladding interface multiple times as it travels along the fiber. The cladding reduces reflection loss to a very low level. Additional sheathing provides strength and protection. Fiber diameters range from 4 to over 1000 μm (1 mm), but a typical fiber has a diameter of 50 to 150 μm. Standard fiber-optic links operate with infrared light (800 to 1500 nm).

5-11   LED and Related Digital Displays

Perhaps the most common digital display is the seven-segment LED display pictured in Fig. 5-21. Each segment is individually illuminated by a single (or sometimes dual) LED with a diffusing bar. One lead is brought out for each segment in a standard display (without a built-in decoder). Most units also contain a decimal point. Any digit from 0 to 9 can be displayed by turning on the proper combination of segments.

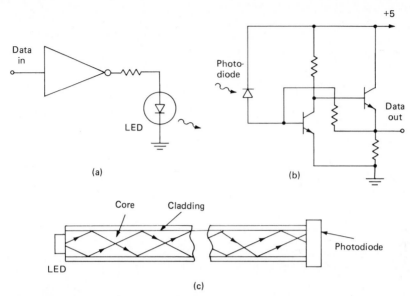

Figure 5-20 Fiber-optic link: (*a*) transmitter; (*b*) receiver; and (*c*) light transmission in a fiber.

Some decoders can display a few other characters, including hexadecimal (0–9, A–F) as indicated in Fig. 5-21*b*.

Although some displays contain built-in decoders (and perhaps latches and counters as well), most require an external decoder/driver. LED displays are divided into two types, common cathode and common anode. Decoders must match the display type. Decoder input is supplied in 4-bit BCD form. The decoder output is seven lines, one for each segment. A transistor driver within the decoder/driver is switched on to turn a segment on. For common-anode displays the anode is connected to the positive power supply (Fig. 5-22*a*). To turn a

Figure 5-21 Digital displays: (*a*) seven-segment array for numerals; (*b*) displayed characters; and (*c*) 18-segment alphanumeric display.

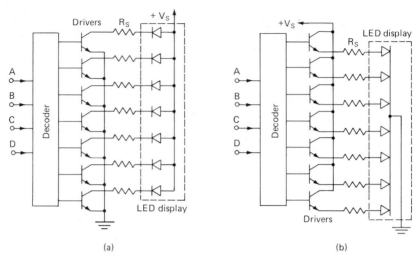

Figure 5-22 Decoder display wiring: (*a*) common anode; (*b*) cathode. Decimal point not shown.

segment on, the cathode is grounded through a current-limiting resistance R_S (typically 100–470 Ω). The open-collector driver transistor is driven to saturation to switch a segment on.

For the common-cathode displays, the display cathode is grounded and the transistor collectors (within the IC) are connected to the positive supply. As with other LEDs, a series resistance must be present on each segment to limit the current to a specified value (typically 20 mA) unless the device has a built-in constant current driver on each segment. A BCD-to-seven-segment decoder/driver and seven resistors are required for each digit with the standard circuit arrangement. Display multiplexing or scanning techniques which require only one decoder are discussed in Chap. 17. Examples of a decoder and display are given in Fig. E-2 at the end of Chap. 3.

The 18-segment display (Fig. 5-21*c*) can outread any alphanumeric character. It is usually driven by a microprocessor without a decoder, although special-purpose drivers are available (e.g., ICM7233). Except for applications requiring a small number of digits (1 to 4), this display is considered inferior to the dot matrix display discussed later (Chap. 17).

The bar display driver (Fig. 5-23) combines a series of comparators with LED drivers. Comparator thresholds are connected to an internal register chain with voltages spaced linearly or logarithmically. This A/D configuration is related to the flash converter (see Fig. 13-17). LED drivers are constant-current and thus no limiting register is needed.

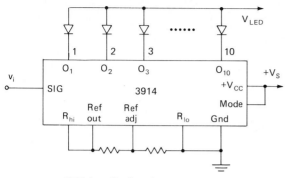

Figure 5-23 LED bar display driver.

5-12 High Current and Voltage Displays

Gas discharge, neon, or incandescent lamp seven-segment displays are similar to the LED displays except for (1) the current and voltage requirements on the drivers and (2) the lamp segments are electrically unpolarized so that either the common-anode or common-cathode drive circuits will work with any display. Neon or gas-discharge displays are bright, large, and low-cost but require a high-voltage drive (about 100 V at 1 mA). Most IC drivers are inadequate, and a high-voltage driver transistor for each segment is a common solution (Fig. 5-24a). A decoder equivalent to that of the LED common-cathode type is required. Segment current is limited by R_S. While the cost of the display is low, the cost of the high-voltage supply and drivers often makes the total cost higher than that of the LED display.

Incandescent displays (Fig. 5-24b) have roughly the same advantages and disadvantages as the gas-discharge displays except that the special power-supply requirement is a relatively high current, usually at a lower voltage. Here also individual segment drivers are usually required. Sometimes a resistor R_W is added to keep the filament warm, thus reducing the warmup shock. These displays require a fair amount of power but can be made as large as desired so that the display can be read at a distance. Direct-view incandescent filament displays (DVD) use the same type of driver.

A vacuum fluorescent display (VFD), a type of cathode ray tube, has a phosphor on a plate which emits light as an electron beam impinges. As with other vacuum tubes, a filament must be heated to produce electron emission (Fig. 5-24c). A potential (15 to 35 V relative to filament) applied to the plate turns on the segment. The entire digit (beam) is turned off by applying a zero or negative voltage to a grid.

While discrete transistor drivers are shown in Fig. 5-24a–c, IC drivers (4, 6, or 7 to a package) are preferable (Fig. 5-24d). Devices

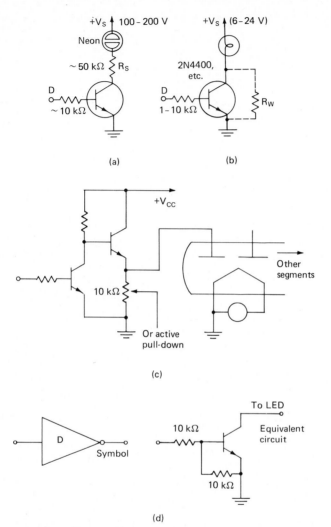

Figure 5-24   Segment drive circuits for (*a*) gas discharge; (*b*) incandescent; (*c*) vacuum fluorescent (VFD); and (*d*) IC driver.

with open-collector outputs (Fig. 5-24*d*) generally can sink more current than comparable devices with open-emitter can source.

5-13   Liquid Crystal Displays (LCDs)

Liquid crystals consist of a high concentration of asymmetric molecules in a transparent organic solvent (Fig. 5-25). These molecules can

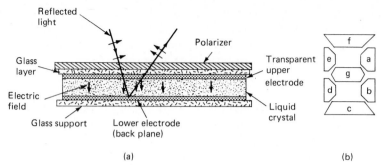

Figure 5-25 Liquid crystal display (LCD) principle (reflection type): (*a*) expanded cross-sectional view; (*b*) top view of five-segment display.

be oriented by an external electric field, which in turn can rotate the plane of polarization of a transmitted beam of polarized light.

A display has an optical polarizer which polarizes the incident optical beam. If the polarized beam is not rotated as it passes through this crystal region, then it will pass through the polarizer again (reflection type) or a similarly oriented polarizer on the other side (transmission type) with little loss. An electric field is applied to the liquid crystal by a voltage (3 to 20 V) across transparent electrodes. The field rotates the plane of polarization by 90° so that the exit polarizer does not allow the light to pass. Thus the area below the electrode (one segment of the display) appears dark (reflection type) in contrast to the white reflected area. Note that the voltage is applied between an individual segment and the common lower electrode or back plane. Either polarity produces the same effect.

A LCD will respond to and may be tested with a dc or a single-polarity (positive or negative) voltage. However, display life is short unless an ac square-wave drive with no average dc component is used. This is accomplished by applying a square wave to the backplane (Fig.

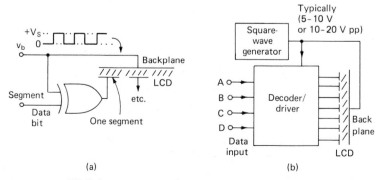

Figure 5-26 LDC driver: (*a*) single segment XOR driver; (*b*) driver display connections.

5-26a). The segment voltage is similarly alternated but is equal to the backplane drive for segment-off and inverted for segment-on. An XOR gate provides the proper segment drive (reversing the truth table) for a data input provided by the standard decoder output. The square wave (30 to 300 Hz) must be symmetrical or the display voltage will have a dc bias.

Decoder/drivers for LCD displays have internal XOR gates (Fig. 5-26b). Normally a decoder/driver is required for each digit, although the backplane is common. Most LCD displays cannot be multiplexed (see Chap. 17 for exceptions).

6

Displacement Sensors

Displacement transducers sense the change in position or displacement of an object. Many sensors are based on measurement of displacement. Linear-displacement transducers sense the displacement Δx or movement along a line. Angular-displacement transducers sense rotation about an axis θ such as the turning of a wheel. Proximity or contactless sensors do not require a mechanical connection to the object being measured. Other sensors may use a shaft to transmit the movement from the object to the sensors, or, like strain gages, be fastened or cemented to the object. For these contact sensors, care must be taken to choose the sensor so that the elastic or frictional force needed to drive the sensor is negligible compared with other forces on the body. In other words, the sensor should not disturb the system being measured. In addition the range of the sensor must roughly match the displacement expected since most sensors become nonlinear or limited beyond some maximum or full-scale value and are insensitive if operated far below this value.

6-1 Strain Gages

A strain gage is intended to measure the small displacements or strains associated with the stress on solid objects such as the bending of a beam. Strain is defined as

$$\varepsilon = \frac{\Delta L}{L} \tag{6-1}$$

where L is the length of an object, or more generally, the distance

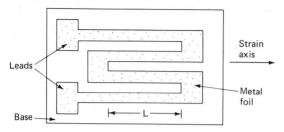

Figure 6-1   Strain gage.

between two reference points fixed in the object. Most gages consist of a metal foil or wire (e.g., Constantan alloy) bonded to a plastic sheet or other insulating base, as illustrated in Fig. 6-1. The base is cemented to the surface of the object under test, and the gage therefore experiences the same strain as this surface. As the length of the foil increases, its electric resistance R increases proportionally, at least for small displacements, as

$$\frac{\Delta R}{R_o} = G_f \varepsilon \tag{6-2}$$

where ΔR is the small resistance increase, R_0 is the unstressed resistance, and G_f is the gage factor.

A resistance increase as the metal strip is stretched occurs both because the strip becomes longer and because its cross-sectional area decreases, according to the relation $R = R_0 + \Delta R = (L_0 + \Delta L)/(A_0 + \Delta A)$, where A_0 is foil-strip cross-sectional area (strip thickness times width) and ρ the resistivity of the conductor. Standard values of R_0 are 120 and 350 Ω. For the ideal case of an incompressible metal $\Delta A/A_0 = -\Delta L/L$, and thus $G_f = 2$.

Another strain gage, the semiconductor piezoresistive type, differs in principle from the metal type in that the resistivity ρ itself is changed by the strain in the crystal. It is produced by heavily doping a surface layer of a silicon crystal. Gage factors of G_f of 30 are typical. The advantage of this high sensitivity is partly offset by a high temperature coefficient. Both the unstrained resistance and gage factor depend on temperature, and some means of temperature compensation is normally required, as discussed below. Predominantly piezoresistive gages are used in IC-type pressure sensors (Sec. 7-2).

Bridge amplifier configurations (Sec. 2-8) are the best readouts for the strain gage (Fig. 6-2a). Good stability is provided by the zeroing potentiometer circuit shown. While the circuit will work with a single gage, often a pair of gages is used in order to compensate for the remaining small temperature sensitivity of the gage.

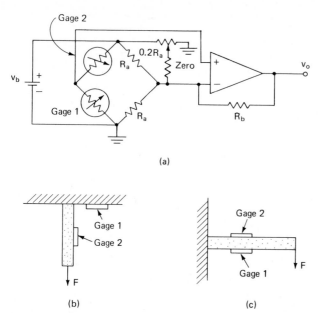

(a)

(b)   (c)

Figure 6-2 Temperature compensation of strain gages: (a) bridge configuration; (b) gage position.

The gages should be mounted close to each other so that they are at the same temperature. Electrically they are connected into the two arms of the bridge circuit such that no bridge unbalance or output is produced if the resistance change in the two gages is identical, as it will be if the ambient temperature only changes. Two methods of mounting may be used. In Fig. 6-2b the second gage is mounted where it will not be strained, but close enough to the first or sensing gage so that it is at the same temperature. If possible, as in Fig. 6-2c, the second gage is mounted in a position where the strain is equal in magnitude but opposite to the sign of the strain in the first gage. In this case not only is temperature compensation provided, but the bridge output voltage is twice that of the single-active-gage circuit.

Often four gages, one in each arm of the bridge, are employed in order to increase the output. The high-frequency response is ordinarily limited by the mass of the body to which the gage is attached.

6-2   Inductive and Electromagnetic Sensors

Inductive (variable-reluctance) transducers utilize the change of inductance of a coil (inductor) in the vicinity of a ferromagnetic material, as indicated in Fig. 6-3. The inductance increases as the material is inserted (Fig. 6-3a) or comes closer (Fig. 6-3b). In order to read out the

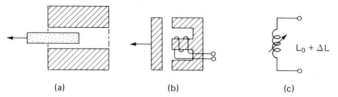

Figure 6-3 Inductive transducers: (*a*) solenoid type; (*b*) air-gap type; (*c*) electrical equivalent.

signal, the inductor is made a part of a bridge (Fig. 6-4) driven by an ac generator. A variable inductor, which is commonly a second (dummy) sensor, is required in one arm of the bridge. Sometimes the second inductor is built into the transducer. The output of the amplifier is an ac voltage proportional to displacement which can be read on an ac voltmeter or converted into direct current by an ac-to-dc converter or phase-sensitive detector

$$v_o = \frac{V_b}{4} \frac{R_b}{R_a} K_L x \tag{6-3}$$

where K_L is the sensitivity (relative inductance change, $\Delta L/L_0$ per unit distance), V_b is the existing voltage, and x the displacement.

A related motion sensor is based on the electromotive force (emf) generated by a coil moving in a magnetic field. One version of the electromagnetic pickup is shown in Fig. 6-5. The output voltage v_e is proportional to the time derivative of displacement dx/dt, more specifically to the velocity of the coil with respect to the magnet (either may move)

$$v_e = K_m \frac{dx}{dt} \tag{6-4}$$

where K_m is a constant for the device that is proportional to the number of turns in the coil and the magnetic field strength. Sensors of

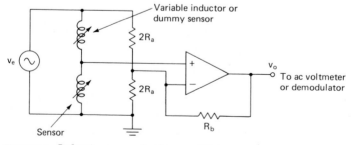

Figure 6-4 Inductive sensor bridge amplifier.

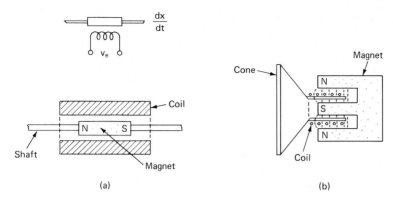

Figure 6-5 Electromagnetic pickups: (a) linear velocity and (b) loudspeaker as a sensor.

the first type are available commercially, or they can be fabricated without difficulty. For some applications, an ordinary loudspeaker can be operated as a motion pickup transducer. An output proportional to displacement x can be obtained by integrating the amplified output of the transducer.

Numerous devices based on the electromagnetic sensor exist, including the dynamic microphone and magnetic phonograph pickup, but because their applications are rather specialized, they will not be covered here.

6-3 Linear Variable Differential Transformer

A linear variable differential transformer (LVDT), Fig. 6-6, is similar in some respects to the inductive transducer, but the principle and electrical characteristics differ. The LVDT has a primary connected to an ac source (typically 60 Hz to 20 kHz) and two secondaries, one on each side of the center, connected so that their output voltages are in opposition. Coupling to the secondaries is determined by the position of the movable magnetic core. When the core is in the center, the coupling is equal, the voltages are equal, and the output voltage is zero. As the core is displaced, coupling to one secondary increases and the output voltage increases. Over a certain range the output is linear with displacement of the core x

$$v_o = v_eKx \tag{6-5}$$

where v_e is the exitation voltage and K is the sensitivity.

In practice an exact null, as measured by a root mean square (rms) meter, is not obtained at center because a small out-of-phase (quadrature) component is present due to capacitive feedthrough and

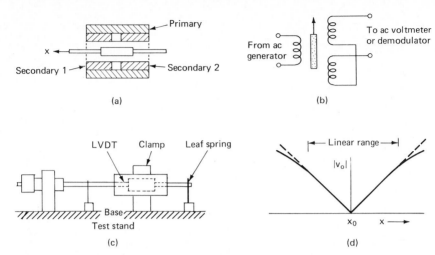

Figure 6-6 LVDT displacement sensor: (*a*) drawing; (*b*) electrical equivalent; (*c*) calibration unit; (*d*) output voltage (magnitude) versus position.

nonideal transfer properties. Further, an rms meter does not register the change in phase of the output as the core moves from one side of center to the other. To overcome these difficulties and to reduce the noise level as well, the use of a phase-sensitive detector (Chap. 14) is desirable. A block diagram of a readout circuit is shown in Fig. 6-7, and a detailed circuit is presented in Example 4.

6-4   Capacitive Transducers

Capacitive displacement transducers are based on the change of capacitance between two electrodes as the separation of the plates changes or the dielectric between them is moved. Three of many versions of this transducer are suggested in Fig. 6-8. Few capacitive transducers are available commercially as separate units because the electrodes are easily incorporated into the body being displaced or mounted on it. Usually one plate is grounded. Capacitances are small, typically 2 to 100 pF, depending of course on electrode area and separation. For a parallel-plate capacitor, the capacitance, neglecting fringing fields, is

$$C = 0.0885 \frac{A}{d} K \tag{6-6}$$

where C is the capacitance in picofacads, pF, A is area in square centimeters, d is electrode separation in centimeters, and K is the dielectric constant ($K = 1$ for air and to 2 to 4 for many plastics). Most capacitance sensors have small capacitances, perhaps 10 to 100 pF.

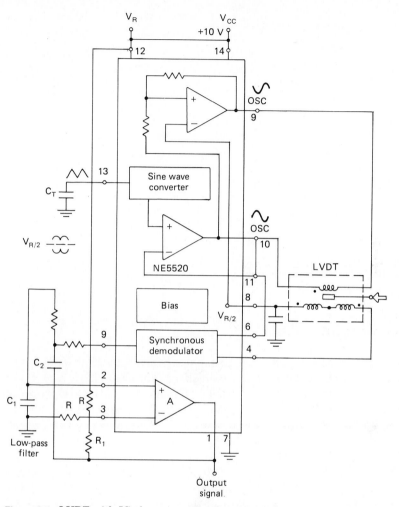

Figure 6-7 LVDT with IC phase-sensitive-detector readout.

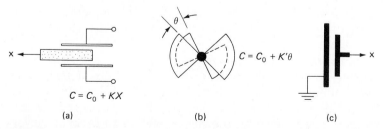

Figure 6-8 Capacitive sensors: (*a*) movable dielectric; (*b*) rotating plated (angular); and (*c*) movable plate.

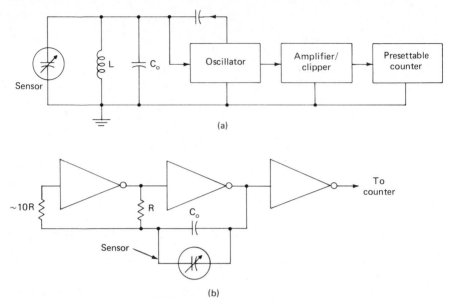

Figure 6-9 Block diagram of frequency shift readouts for a capacitive sensor: (a) LC oscillator; (b) relaxation oscillator.

Advantages of the capacitance transducers are their high sensitivity and simple noncontact (electrode) mounting. Nonlinearity is a problem unless careful attention is paid to the electrode geometry and fringing fields. Since the capacitance change is small, the connecting cables are a problem, as is the requirement of a relatively complex readout, which is also a consequence of the low capacitance.

A bridge amplifier with FET op-amp, similar to that shown in Fig. 6-4, may be employed as a readout. The inductors are replaced by the transducer and an adjustable capacitor. A high-value resistor across each capacitor provides the required dc current return to ground for the op-amp input. Care must be taken in the bridge wiring to minimize stray capacitance, especially changes in stray capacitance, which are equivalent to a drift or noise at the output. Rigid wiring and shielding, as well as mounting of the bridge and amplifier very close to the electrodes, is necessary.

A second type of readout is based on the frequency shift of an oscillator circuit in which the capacitive transducer is a frequency-determining element. In Fig. 6-9a, an LC oscillator (Chap. 11) version is given. Although a wide range of operating frequencies is practical, commonly a center frequency in the region of 1 to 10 MHz is chosen. Calculation of the frequency shift due to changes in the transducer capacitance ΔC is straightforward. When ΔC is small, the following linear term is obtained by binominal expansion:

$$\frac{\Delta f}{f_0} = \frac{-\Delta C}{2C_0}\tag{6-7}$$

where $f_0 = 2\pi LC_0$ and C_0 is the sum of the stray capacitance and the minimum capacitance of the transducer. Note that the frequency range of sensitivity is determined by the choice of C_0 and L. If a frequency meter is used to measure the output, it is convenient to adjust L and C_0 so that the frequency shift for a unit displacement is an integral amount. A counter which can be preset to a negative number is suggested so that the output display can be set to zero when $\Delta C = 0$. Sometimes it is possible instead to adjust the count so that the initial frequency corresponds to a higher digit which is not displayed.

For low-cost applications, the dual-inverter CMOS relaxation oscillator (Fig. 6-9b) is especially well suited (see Sec. 11-4). Accuracy is limited because the frequency depends on the inverter logic threshold, which varies in temperature and perhaps drift with time. Better stability is achieved by a standard timer-type multivibrator (such as the 555), but the sensitivity is less since these devices are not intended to work with the very low capacitance changes usually encountered (1 to 0.01 pF). Typical operating frequencies are 30 kHz to 1 MHz.

6-5 Digital Optical Displacement Transducers

When relatively large displacements are to be determined, a digital encoded transducer is often the best choice, especially if the readout is digital. Two types of transducers will be discussed, incremental and absolute. In Fig. 6-10 the incremental type of pickup is illustrated for the range subdivided into 10 intervals. For many practical applications, subdivision into more intervals, perhaps 100 or 256, would be required, while for others one pulse per turn (half opaque, half transparent) would do. Resolutions to 16 bits better are available. As the wheel rotates, light pulses converted by the light detector into

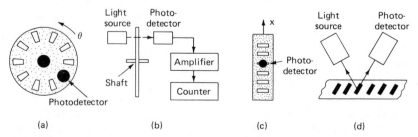

Figure 6-10 Incremental digital displacement sensors. Angular transmission type: (a) front view and (b) backside. Linear reflection type: (c) front view and (d) reflection-type side view.

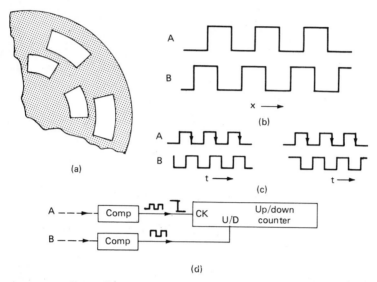

Figure 6-11 Reversible incremental position detector. (*a*) Diagram of encoding wheel; (*b*) channel outputs as a function of position; (*c*) channel outputs as a function of time; and (*d*) block diagram of counter circuit.

electric pulses are amplified and counted by a decade counter. The digital reading of the counter is thus equal to the angular displacement of the wheel and/or number of revolutions. Linear transducers can be constructed in a similar fashion.

It should be noted that the frequency of the pulse train at the output of the comparator is proportional to velocity. A frequency counter connected to the output in place of the pulse counter will register velocity directly.

Reversible displacement can be measured with a two-channel encoder and an up-down counter, as illustrated in Fig. 6-11. The two channels are displaced by 90° in position (Fig. 6-11*b*). One channel (A) is the edge-sensitive clock input, while the other (B) determines the direction of the count. As the timing diagram of Fig. 6-11*c* indicates, the *falling* edge of the clock channel occurs at different phases of the up-down channel for the two directions of motion and this difference determines the counter direction.

Absolute digital displacement transducers, both angular and linear, use several optical pickups to provide an encoded output directly, as illustrated in Fig. 6-12. Transmission types (opaque-clear) are most popular. Each bit requires a separate photodiode pickup as well as a separate strip on the encoding card. With this arrangement there is no ambiguity with respect to card position or direction of movement. If BCD coding is used, the outputs can be sent directly to a BCD-to-seven-

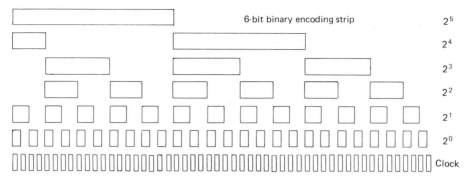

Figure 6-12 Absolute digital displacement encoders; a 6-bit encoder with clock channel is shown.

segment decoder and display. The main disadvantage is the multiple pickups required.

Extension to a larger number of bits is straightforward. In order to remove ambiguity of the detector outputs at transition points (e.g., changing from binary **011111** to **100000**), a clock channel is usually added. Alternatively, a special code (Gray code), where only one channel changes state at each increment, is sometimes employed. Resolutions of 8 to 10 bits are practical with plastic-base encoders (15 bits or more with glass-base encoders).

Generally, angular encoding disks are capable of higher accuracy than linear encoding strips because the temperature expansion of the disk affects the accuracy very little. As the resolution becomes finer, the optical system and photodetector array geometry become more critical, as one might expect.

Measurement of small displacements (along one direction) is made easier by employing a Moire pattern on a pair of movable strips. Both strips have a parallel bar pattern (Fig. 6-13a, b) which consists of

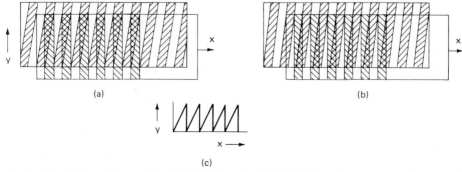

Figure 6-13 Moire pattern strips for detection of small displacements.

alternating opaque and transparent areas, but one set of bars is inclined at a small angle with respect to those on the other strip. Transparent spots occur where the clear bars overlap. These transparent spots move up (y direction) when one strip is moved laterally (x direction) with respect to the other, as indicated by Fig. 6-13c. If the bar pattern has only a slight angular tilt, the spot displacement up is 10 or 100 times the displacement across, thus providing a magnification of movement which is much easier to detect by an array of optical pickups.

6-6   Piezoelectric Transducers

Piezoelectric transducers take advantage of the voltage developed across certain crystals when they are strained. Ionic crystals within which the repeating arrangement of atoms (unit cell) is asymmetric, develop a separation of charge across the crystal faces because the positive and negative ions separate slightly when the crystal is strained. Two common piezoelectric crystals are quartz and barium titanate. Electrets, including the polymer polyvylidine fluoride, also have excellent piezoelectric properties. The magnitude of the voltage and the direction of strain sensitivity depend on the cut of the crystal with respect to the crystallographic axis (or the polarization direction in the case of electrets), but the response is always of the form

$$v_x = K_x \varepsilon \tag{6-8}$$

where v_x is the voltage developed across crystal, K_x is sensitivity, and ε is the strain.

Because the crystal is a nonconductor, the generator voltage is coupled to the electrodes at the crystal faces via the crystal capacitance C_x, as indicated in Fig. 6-14. Since any practical amplifier has a finite input resistance R_L, the result is that the voltage v_x must pass through a high-pass network (Chap. 10), where $f_x = \frac{1}{2}\pi R_L C_x$, and thus the transducer will not respond to direct current. A high-input-impedance FET op-amp is essential at lower frequencies, although at high frequencies lower-impedance amplifiers may be adequate. Each crystal has one or more sharp mechanical resonance points (22 kHz to 10 MHz), at which point the output is several orders of magnitude higher than off resonance (see Sec. 11-7). Alternatively, the sensitivity to strain can be expressed in terms of the charge Q_x developed across the crystal face.

$$Q_x = K_{q\varepsilon} \varepsilon \tag{6-9}$$

where the transducer charge and voltage are related through capaci-

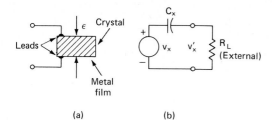

(a)   (b)

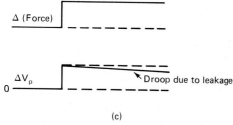

(c)

Figure 6-14 Piezoelectric transducers: (a) geometry; (b) equivalent circuit; (c) response.

tance $C_x = Q_x/v_x$. Still other alternatives are to express the sensitivity in terms of applied force or pressure (Chap. 7).

Voltage-mode readout consists of a very high input impedance (MOSFET or maybe JFET) operational amplifier (Fig. 6-15a, b). Adequate output voltage is provided by a unity-gain amplifier in most cases and, in fact, the voltage can be so high that input pulse protection on the op-amp input is necessary since taps or sudden impulses can result in high-voltage pulses. A high-value resistor R_L in Fig. 6-15a provides the necessary dc current path from the op-amp input to ground; the circuit time constant is thus $R_L C_x$. In the switched-input circuit (Fig. 6-15b), the transducer is kept grounded by the FET switch until it is time to read the strain (actually the difference in strain between the off and on condition). The read time should be brief.

Charge-mode readouts (Fig. 6-15c) provide an output voltage proportional to the change in charge Q_x.

$$V_o = \frac{Q_x}{C_f} = \left(\frac{K_{q\varepsilon}}{C_f}\right)\varepsilon = K_{\mu\varepsilon}\left(\frac{C_x}{C_f}\right) \tag{6-10}$$

It is assumed that the input change is rapid ($<<R_L C_f$). This equation is equivalent to that for the standard inverting amplifier with the

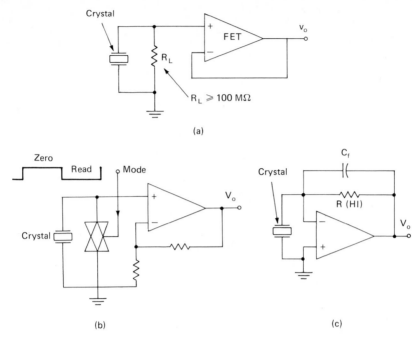

Figure 6-15 Piezoelectric transducer amplifiers: (*a*) voltage mode (*R* input); (*b*) voltage mode (switched input); and (*c*) charge mode.

resistances replaced by the impedances ($Z_c = -j/\omega C$). An advantage of this charge amplifier is that it is fairly insensitive to leakage across the transducer since the op-amp holds the transducer voltage to nearly zero.

6-7   Resolvers and Inductosyn Transducers

A small motor-generator pair, the resolver and transmitter, have been specialized for the sensing of angular position. Two stators fixed at angles of 90° are driven by a sinusoidal current (typically 400 Hz) displaced by 90° in phase (Fig. 6-16). As the rotator is rotated, the phase of the induced signal varies with the angular position and, in fact, the phase angle and rotation angle can be made equal. Readout of the phase requires a digital phase meter or phase-sensitive detector (Chap. 14). The three-phase (three-stator) versions are termed syncros. The linear version, termed inductosyns (Fig. 6-16*b*), consists of an accurately positioned fixed trace and a pair of movable traces (windings). The two windings are displaced by $\frac{1}{4}$ turn and, correspondingly, the small induced voltages are displaced by 90°. By measuring the number and phase of the cycles, the displacement of the movable section can be accurately determined.

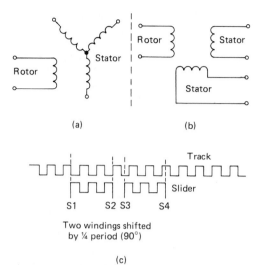

(a)　　　　　　　(b)

Track

Slider

S1   S2 S3   S4

Two windings shifted
by ¼ period (90°)

(c)

Figure 6-16 Motor style position transducers: (a) resolver and (b) inductosyn.

6-8 Hall Magnetic Sensors and Proximity Detectors

A more sensitive static magnetic detector, and one which in some devices exhibits proportional response, is the Hall sensor. It is based on the Hall effect, which all semiconductors possess to a greater or lesser degree. Current i_c is passed through the semiconductor (Fig. 6-17a) in a direction perpendicular to the magnetic field (at least a component must be perpendicular). Voltage leads are placed perpendicular to the current flow direction, as close as possible on an equipotential line when the magnetic field B is zero, so that $\Delta v = 0$ for $B = 0$. For maximum response, the plane of the voltage and current, i.e., surface of the semiconductor slab, must be perpendicular to the magnetic field. When the field is applied, the charge carriers giving rise to the current flow are deflected slightly by the magnetic field and drift from one voltage electrode to the other (depending on current and magnetic field direction). The net result is that the voltage Δv between the pickup arms is proportional to B, the strength of the magnetic field at the semiconductor. Linear sensors produce an output voltage proportional to the field strength (Fig. 6-17b). With careful temperature compensation, sensitivities of better than 1 G are possible. Uncompensated units have zero drifts up to 100 G, and, therefore, threshold levels of 200 to 500 G are generally chosen for proximity detector applications. Small magnets (3 mm diameter, 10 mm long) can produce fields of this magnitude at distances of 1 to 5 mm.

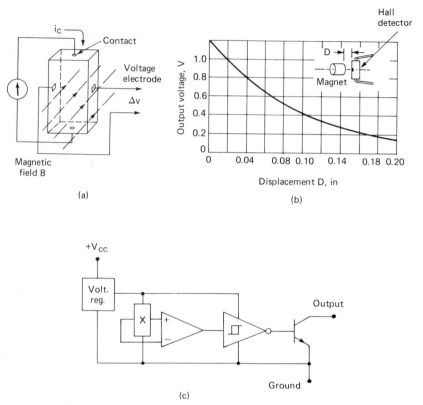

Figure 6-17 Hall-effect magnetic sensors: (*a*) principle; (*b*) characteristics of linear type; (*c*) block diagram of digital type.

A Hall proximity detector, an integrated circuit (Fig. 6-17*c*), employs a comparator with TTL-compatible output to detect the threshold level of a magnetic field. As a replacement for a limit switch, it offers the advantages of higher reliability, noncontact detection, and lower cost. Several modes of proximity detection are indicated in Fig. 6-18.

In addition to the Hall-effect type, several other proximity detectors are used industrially. Magnetic reeds (Fig. 6-19*a*) deflected by a nearby magnet are a low-cost method. Radiofrequency (40 to 200 KHz) loss types detect proximity by loading of an inductor in an oscillator circuit due to eddy current losses when a conducting metal slab approaches. The oscillator is designed so the rf amplitude reduces substantially with LC resonance circuit loading, and this voltage drop is detected by a comparator with a TTL-compatible output.

Detection is based on capacitance (Sec. 6-4) and has the advantage of being sensitive to nonconductors. Another detector is based on the reflection of a focused light beam (Chap. 5).

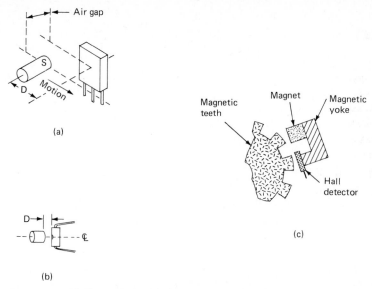

Figure 6-18 Hall proximity detectors: (a) end-on magnet motion; (b) slide-by magnet; (c) fixed sensor, variable reluctance.

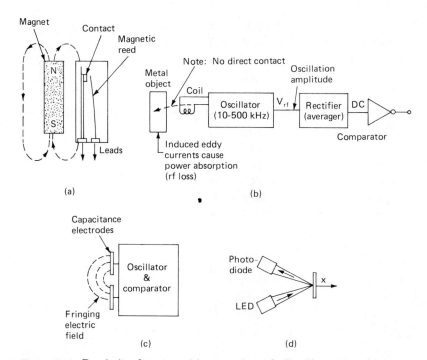

Figure 6-19 Proximity detectors: (a) magnetic reed; (b) rf loss; (c) capacitance; and (d) optical.

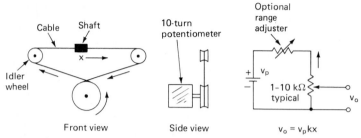

Figure 6-20 Potentiometers as a linear displacement sensor: (*a*) linear not on; (*b*) angular type; (*c*) readout circuit.

6-9   Other Displacement Transducers

Potentiometers are convenient displacement transducers in applications where the movement is of large amplitude, is slow, and friction is not a problem. A 10-turn potentiometer is best because of its good electrical linearity and resolution. A servo type with ball bearings may be required for heavy use. A friction (clutch) connection is usually needed to prevent the potentiometer from being driven beyond its range. Its application as an angular transducer (limited to 10 revolutions) is obvious and, as Fig. 6-20 indicates, the extension to linear displacement measurements requires only the addition of a simple cable drive. Readout in terms of a voltage proportional to displacement requires that a fixed voltage be applied across the potentiometer. Where low-output impedance is required, a unity-gain amplifier can be added.

Resolution of the potentiometer is typically limited to 0.02 to 0.5 percent of full scale because of its mechanical construction and is not suited for measuring small displacements. This problem and the

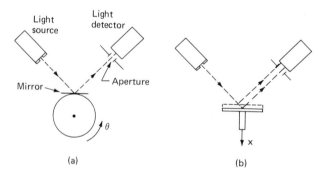

Figure 6-21   Optical lever displacement detector: (*a*) angular; (*b*) linear.

possible mechanical difficulties (friction, backlash) are often overlooked by those who are focusing on the selection of a displacement transducer with the simplest electrical readout.

Optical levers consisting of a light source, reflecting surface, and photodetector can be used to measure small displacements. Several arrangements are illustrated in Fig. 6-21. Displacement is detected by the movement of the edge of the light beam across the photodetector aperture, which causes a variation in detected intensity. Since the light intensity does not fall off linearly with distance from the center of the beam, the overall response of the transducer is quite nonlinear except for very small displacements, and the response range is also quite limited. These transducers are most useful in detecting small vibrations or as a displacement null detector.

Chapter

7

Pressure and Force Sensors

Pressure and force transducers (load cells) consist of an elastic element and an attached displacement sensor. The applied force produces a small but reproducible strain or displacment of the diaphragm or other elastic element. Most readout circuits are similar or identical to standalone displacement sensors and discussion will not be repeated here. Output is calibrated in the desired pressure or force units. The discussion here focuses on the electrical rather than the equally important mechanical aspects of electronic sensor design.

Pressure sensors are divided into absolute, gage, or differential types, depending on whether the opposite side of the diagram is a closed container (perhaps under vacuum), left open to the atmosphere, or connected to a second inlet to allow measurement of pressure differences.

7-1 Diaphragm-Type Pressure Sensors

Most pressure transducers utilize diaphragms as elastic elements since they show an especially linear and reproducible response and are relatively easy to manufacture. Displacement is small, however, and therefore sensitive sensors, such as capacitance or strain gages, are required. Two methods of mounting are shown in Fig. 7-1a, b. The clamped-edge type is more difficult to manufacture and has less displacement but is more reproducible. When capacitance sensing is used, the metal diaphragm is one plate of the capacitor (Fig. 7-1c). A pair of capacitance plates, connected in an ac bridge arrangement, is used by the manufacturers of the most accurate and sensitive (and costly) pressure sensors available.

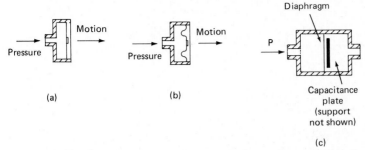

Figure 7-1 Diaphragm pressure sensors: (*a*) taut, edge clamped; (*b*) corrugated; (*c*) integral capacitive sensor.

Strain gages are the most common means of measuring diaphragm deflection. Nearly always four gages are employed in a bridge configuration to increase output and, especially, to reduce the temperature coefficient. Two of the gages must be placed in a region of compression and the other two in a region of tension, as indicated by Fig. 7-2. The bridge readout (Fig. 7-3*a*) is similar to that previously discussed (Sec. 4-8 and 6-1) except that a small variable resistor may be added in series with the bridge power supply to allow for calibration. Calibration is required because the diaphragm fabrication and gage mounting are difficult to control precisely. A known pressure (or force) is needed for calibration.

Additional temperature compensation of the zero and full scale may be required, in part because the elastic element is somewhat temperature sensitive. One method is to add temperature-sensitive resistors (R_{CTC}, R_{ZTC}) to the bridge circuit (Fig. 7-3*b*). Because these trim resistors cannot be adjusted independently, it may be desirable to

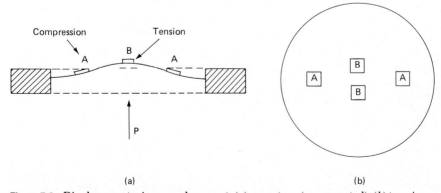

Figure 7-2 Diaphragm strain gage placement: (*a*) curvature (exaggerated); (*b*) top view.

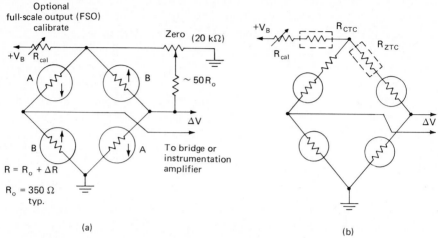

(a)   (b)

Figure 7-3 Bridge for strain gage pressure and force sensors: (*a*) without extra temperature compensation; (*b*) with full temperature compensation.

introduce the compensation into the offset and gain resistors of a subsequent stage or of the bridge voltage regulator.

7-2   Silicon Pressure Transducers

Solid-state sensors employ silicon both as a piezoresistive strain gage and as the elastic diaphragm. Normally the gage and diaphragm are part of the same chip (a good example of an IC-type sensor) although the strain gage chip can be obtained separately (Fig. 7-4*a*). The gage

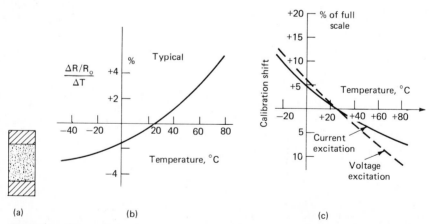

(a)   (b)   (c)

Figure 7-4 Piezoresistive strain gages: (*a*) drawing of gage; (*b*) variation of gage factor and resistance with temperature; (*c*) temperature response in bridge configuration.

factor G_f and thus sensitivity of a piezoresistive-type strain gage is much higher than that of metal gages (Fig. 7-4b) because the resistivity of the gage is changed by the strain (not just the dimensions, see Sec. 6-1). However, the temperature coefficient is also very high and temperature compensation is a major design problem. As with other diaphragm-type sensors, the gages are used in sets of four in a bridge configuration to reduce greatly the change in resistance with temperature. Still substantial errors remain because (1) the gages are not perfectly matched, causing a zero shift, and (2) the gage factor G_f is temperature-sensitive (Fig. 7-4c). While current excitation is sometimes used to reduce the temperature coefficient of the (full-scale) output (by a factor of 2) better results are achieved by compensating the bridge voltage or amplifier gain. Zero drift compensation can be achieved by shunting a bridge element (which changes with temperature) by a high-value trim resistor (which does not change with temperature).

Silicon semiconductor pressure sensors have the diaphragm etched into the silicon chip (Fig. 7-5). The strain gage is placed (by diffusion of doping compounds) onto the opposite surface. Different manufacturers etch with different patterns intended to produce the maximum strain and thus sensitivity. Silicon pressure sensors (diaphragm and

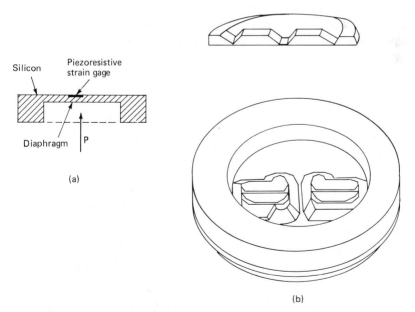

Figure 7-5 Silicon diaphragm pressure sensors: (a) flat etched diaphragm; (b) shaped diaphragm for higher sensitivity.

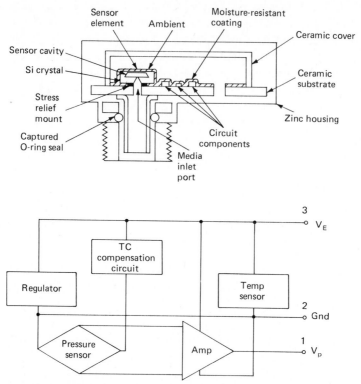

Figure 7-6 Semiconductor pressure sensor with internal amplifier: (*a*) sensor; (*b*) amplifier block diagram. Sensym is one manufacturer.

bridge) are often combined with the readout circuit (Fig. 7-6). Temperature compensation and linearization are included in the circuit.

Capacitance electrodes may be used instead of strain gages on silicon diaphragms. Pulses with a frequency proportional to pressure are produced by the readout circuit.

7-3 Other Types of Pressure Gages

Numerous methods of converting pressure into a sizable displacement have been derived, a few of which are shown in Fig. 7-7. Displacements are measured with a LVDT, a variable-reluctance sensor, or perhaps simply by a potentiometer. While large displacements are produced by these elastic elements (Bourdon tube or bellows), the reproducibility is inferior to diaphragms.

Dynamic pressure (pressure changes or sound) may be measured by

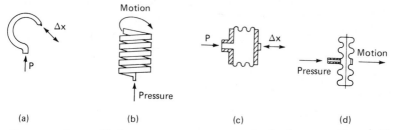

Figure 7-7 Large displacement pressure gage elastic elements: (*a*) and (*b*) Bourdon tubes; (*c*) bellows; (*d*) capsule.

piezoelectric transducers which are similar to the force transducers described below.

7-4 Load Cells

Load cells or force transducers are similar to pressure sensors in that an elastic element converts the applied force to a displacement which is then measured with a displacement transducer. Strain gage bridges are also used extensively and the readouts, calibration, and temperature compensation techniques are nearly identical. A number of configurations are shown in Fig. 7-8.

Piezoelectric crystals are good sensors for transient (not steady) forces (Fig. 7-9). As indicated in the previous discussion of piezoelectric sensors (Sec. 6-6), they can be equally well specified in terms of force (or pressure) as well as strain (or acceleration). They have a very fast response, limited only by the crystal mass, but cannot be used to measure static dc forces. With high-input-resistance amplifiers now available (Fig. 6-15), even rather slow transients can be handled (Fig. 7-9*b*), and this limitation is not severe.

7-5 Specification of Sensor Errors and Nonlinearity

Pressure and force sensors, like all sensors or indeed all electronic devices, are not perfectly linear or reproducible. Testing and calibration of sensors, however, is more difficult than testing of many other electronic devices. A standard for the physical variable being measured (e.g., pressure) is required in addition to the readily available electrical standard (e.g., voltage). For this reason, sensor users usually must rely on manufacturers for testing and therefore should understand the meaning of manufacturers' specifications (as defined, for example, by the Instrument Society of America).

Most sensors are specified in terms of a linear response (dashed line

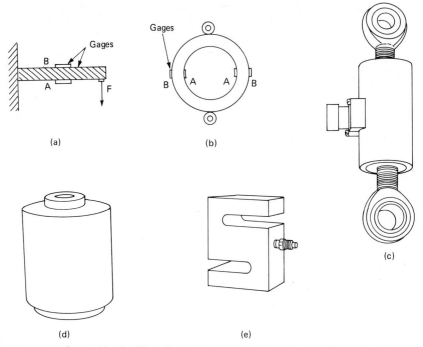

Figure 7-8 Several load cell configurations: (*a*) cantilever beam; (*b*) proving ring; (*c*) Shear beam; (*d*) standard housing.

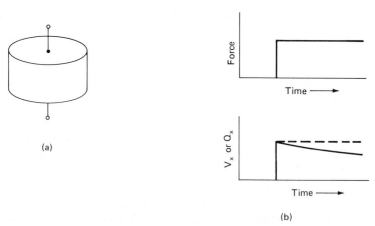

Figure 7-9 Piezoelectric force transducer: (*a*) geometry; (*b*) transient response.

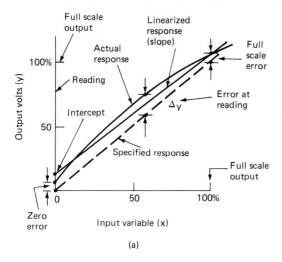

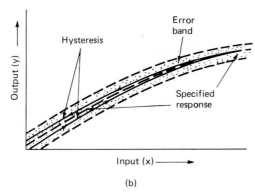

(b)

Figure 7-10 General sensor response: (*a*) deviations from linear; (*b*) comformability definitions.

in Fig. 7-10) passing through zero. Actual response may be slightly nonlinear, have a small zero offset, and not quite match the specified full-scale output. A close approximation to the linearized response or a (least-squares) best-fit straight line is often calculated. Sometimes the line is fixed at zero and in this case the intercept is the zero error. When a zero offset adjustment can be made, the straight line may be set to pass through the actual response intercept. Sometimes only the best fit over the response as a whole is taken without special attention to the zero. In any case, the actual response can be compared against the selected, specified straight-line reponse at the three points: (1) zero,

(2) full-scale output, and (3) reading (desired data measurement point). Errors (differences between actual response and specified responses at any point) are specified as a percent of full scale or as a percent of reading. If the manufacturer specifies the full-scale (calibration or slope) error and zero (intercept) error as a percent of full scale, the user can calculate the error at the desired point as a percent of reading, provided the response is close to linear.

Pressure, force, and many other sensors also have some degree of hysteresis error (Fig. 7-10b), defined as the difference between output response for an increasing and decreasing variable taken at the same value of that input. The error at any particular reading should not be greater than that specified by the error band for any condition (a conservative method of specification).

8

Flowmeters and Fluid Measurement Sensors

The more popular numerous gas and liquid flow rate and velocity transducers are described here. Selected liquid-level gages as well as humidity and moisture sensors are also reviewed.

8-1 Flowmeters Based on Pressure Drop

The turbine (Fig. 8-1) under ideal conditions is a highly accurate flowmeter with a rotation rate proportional to fluid velocity. An electromagnetic pickup or proximity device placed at one side produces a pulse as the turbine blade passes the sensor. The amplified pulse output is proportional to fluid velocity. Clean, noncorrosive fluids are necessary.

Conversion of the flow units from fluid velocity u to volumetric flow rate (in units of liters per second, for example) requires a knowledge of the pipe diameter and the relative velocity distribution across the pipe diameter. Actually the force on the turbine blades, and thus the rotation speed, is in proportion to the momentum (and thus mass) so that the turbine is classed as a mass flowmeter. In most cases, the mathematical relation is complicated and the calibration is supplied by the manufacturer.

A related, but less accurate, device is the paddle-wheel type. Its calibration is highly dependent on fluid viscosity and pipe size, but it can be constructed from materials which are more corrosion resistant than the turbine meter.

Orifice flowmeters rely on the pressure drop across an orifice plate in

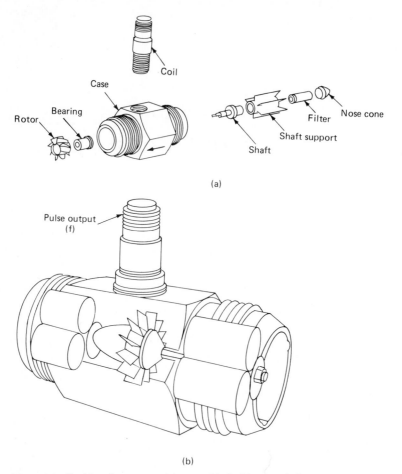

Figure 8-1   Turbine flowmeter: (*a*) assembled; (*b*) expanded view.

pipe containing the fluid (Fig. 8-2*a*). Differential pressure transducers (Chap. 7) measure the pressure drop ΔP. The volumetric flow rate Q_f is nonlinearly related to the pressure:

$$Q_f = K_f \beta A^2 \sqrt{\frac{\Delta P}{\rho}} \tag{8-1}$$

where A is the pipe diameter, β is the ratio of orifice to pipe diameters (typically 0.2 to 0.5), ρ is the density, and K_f is a constant which depends on the gas (or liquid) type. For air (at 1 atm) at a flow rate of 1 L/s, ΔP will be 1 torr for $\beta = 0.3$ and $A = 5$ cm. While this is a popular flow gage because of the simple mechanical construction, the nonlinear and fluid-dependent response is a disadvantage.

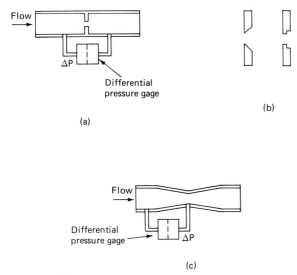

Figure 8-2   Pressure drop flowmeters: (*a*) orifice type; (*b*) orifice detail; (*c*) venturi type.

The venturi tube reduction in pressure in the throat (Fig. 8-2*c*) utilizes the same effect as the lift of an airplane wing. Flow rate also varies with the square root of pressure difference. Its high cost limits its use to applications where minimum pipe constriction is necessary (high-velocity air or liquid).

Water flow in large, open channels may be determined by measuring the pressure head (or height of the head) as the water flows over a small dam or weir, often V-shaped at the center. Again, flow rate varies approximately as the square root of the pressure difference.

An interesting method of measuring fluid velocity is based on the frequency of vortex production as a streamlined flow strikes a nonstreamlined obstacle (Fig. 8-3). Vortices are produced alternately on either side of the obstacle at a periodic rate (the same cause as a flag flapping in a wind). Correspondingly, an alternating pressure difference occurs between the two sides. Above a threshold (below which no vortex production occurs), the frequency is proportional to fluid velocity (Fig. 8-3*b*). Of many methods of detecting the ac pressure difference, a hot thermistor placed in a small channel between the two sides is illustrated here. The alternating directions of flow through the channel periodically cool the self-heated thermistor (see next section), thus producing an ac signal and pulses (Fig. 8-3*c*) at twice the vortex frequency.

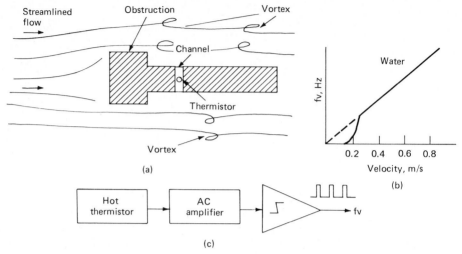

Figure 8-3 Vortex shedding flowmeter: (*a*) flowmeter geometry; (*b*) response; (*c*) readout block diagram.

8-2 Other Flowmeters

The cooling effect of a flowing fluid is employed in the heated thermosensor flowmeter. In one version (an update of the hot-wire anemometer) a thermistor is self-heated by operating the bridge at a higher voltage. The hot thermistor may be 30°C above the ambient reference (heat-sink) thermistor temperature when the fluid velocity is zero. The temperature drop ($\Delta T_u = T_{u=0} - T_u$) is approximated by

$$\Delta T_u = K_u P_w u \tag{8-2}$$

where P_w is the input power and K_u depends on sensor geometry and construction. It is most suitable for low-cost and accurate gas flow measurement applications, such as building ventilation tests. Better accuracy is achieved by separating the heater and thermosensor (Fig. 8-4*c*), in which case the temperature rise is held constant by controlling and reading out the heater power (P_h).

A more accurate type of heat flux flowmeter is shown in Fig. 8-4*d*. Usually it is used with gases and termed a mass flowmeter* because the heat capacity of a gas increases with its molecular weight and the cooling effect is proportional to the mass of the gas rather than the volume.

---

* It should be noted that mass flowmeters may be calibrated in volumetric units (e.g., liters per second) and volumetric flowmeters in mass units (e.g., grams per second) for a specific fluid so that the displayed unit is not a good indication of the type of flowmeter.

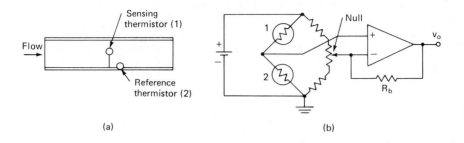

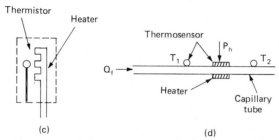

(c)   (d)

Figure 8-4 Thermal-type mass flowmeters: (*a*) internally heated thermistor; (*b*) readout; (*c*) separated heater type; (*d*) heated capillary type.

Gas (and sometimes liquid) flowing through a long, thin tube is heated electrically and the temperature rise $\Delta T_h = T_1 - T_2$ monitored (Fig. 8-4*d*). A control adjusts the heater power P_h such that ΔT_h is constant (typically 8 to 40°C). Under these conditions, the flow rate is given by

$$Q_f = K_f\ \Delta T_h P_h \qquad\qquad (8\text{-}3)$$

where K_f depends on tube diameter and gas heat capacity. Accuracy is best for thin capillary tubes, in which case applications are limited to clean (but perhaps corrosive) gases.

Flow velocity of conducting liquids may be accurately measured by an electromagnetic flowmeter (Fig. 8-5). A nonconducting pipe is placed within a magnetic field. Charge carriers moving with the liquid are deflected by the magnetic field and collect along the side of the pipe (D = diameter). The charges (q) produce an electric field $E = \Delta V_e/D$ and force (qE) which at equilibrium balances the force Bqu produced by the magnetic field (B), leading to the relation

$$\Delta V_e = BDu \qquad\qquad (8\text{-}4)$$

where u is the (average) velocity of the fluid. Usually, ΔV_e is small (0.1

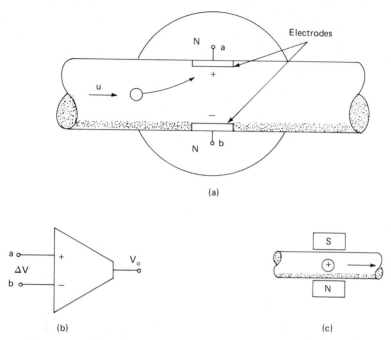

Figure 8-5 Electromagnetic flowmeter: (*a*) top view; (*b*) side view; (*c*) block diagram of readout.

to 10 mV), comparable to contact potentials caused by impurity absorption at the electrodes, so that ac excitation and amplification are necessary. Unfortunately, production of an ac magnetic field is much more difficult than simply employing a permanent magnet. Accuracy and speed of response are good unless the conductivity becomes low (under 10^{-5} S, comparable with distilled water). It should be noted from Eq. (8-4) that the voltage output does not depend directly on the conductivity, although speed of response (and leakage resistance error) does.

Doppler flowmeters rely on the change in frequency (Δf) of sound wave reflected from a moving fluid (Fig. 8-6*a*). The Doppler shift is given by

$$\frac{\Delta f}{f_o} = \frac{2u \cos \theta}{C} \tag{8-5}$$

where f_o is the transmitted frequency, c is the sound velocity, u is the liquid velocity at the point of reflection, and θ is the angle between the fluid and sound velocities assumed here to be the same for transmitted

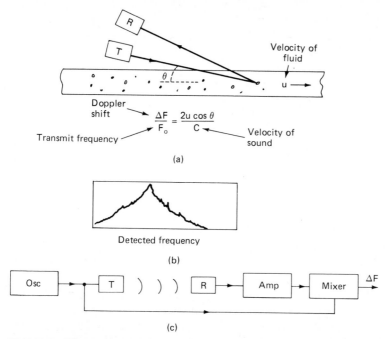

$$\frac{\Delta F}{F_o} = \frac{2u\cos\theta}{C}$$

Figure 8-6 Ultrasonic (Doppler) flowmeter: (a) geometry; (b) response; (c) block diagram of readout.

and received waves. Sound waves are reflected only from discontinuities (in accoustical impedance) such as small bubbles or even density fluctuations due to turbulence. Unless the sound is focused and/or the fluid velocity is uniform across the pipe, the reflected wave will contain a distribution of velocities (Fig. 8-6b). Separation of the small Doppler signal from the transmitted frequency is not difficult and, in fact, can be made part of the heterodyne detection process where a portion of the transmitted frequency signal is mixed with the received signal to obtain a difference or beat frequency (Fig. 8-6c). A high-pass filter retrieves the dc component from the desired ac Doppler signal. The readout circuitry must convert the weighted average of the Doppler signal frequency spectrum to the desired velocity or flow rate units. Because of the difficulty in obtaining a proper average (and perhaps also uncertainty due to refraction of the beam through the pipe wall), commercial flowmeters require a calibration adjustment for specific installations. A related type of flowmeter depends on the difference of transmission times (or phases) of sound waves traveling with and against the flow.

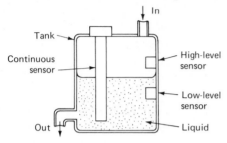

Figure 8-7 Liquid-level sensors.

8-3 Liquid-Level Sensors

Liquid-level gages measure the height of a liquid in a container. They can be classfied as discrete and continuous (Fig. 8-7). A discrete level sensor simply indicates whether the liquid is above or below the position of the sensor, while a continuous type measures the depth of the liquid. Discrete sensors are often used in pairs in high- and low-level indications which use control valves or pumps to maintain the liquid level between these two positions. Because many physical properties of a liquid differ from the air above, a wide variety of sensors are possible, and the choice depends on the particular liquid and size of container (different sensors would be appropriate for sulfuric acid in a chemical plant, molasses in a storage tank, liquid nitrogen in a physics experiment, and the level of Lake Erie).

Several discrete level sensors are illustrated in Fig. 8-8. An internally heated thermistor (or a thermocouple) mounted on a heater mounted to the container wall at the desired level is shown in Fig. 8-8a. Because the thermal conductivity of the liquid is better than a gas, the thermistor will be much cooler when immersed in the liquid. The principle is the same as used in fluid-flow and vacuum gages, discussed above. Readout circuits are given in Chap. 4.

The second sensor (Fig. 8-8b) is a piezoelectric crystal which is made

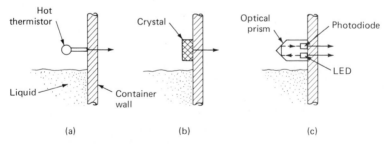

Figure 8-8 Discrete liquid-level sensors: (a) thermoconductive; (b) oscillating crystal; and (c) optical reflection.

a part of an oscillator. When the crystal is immersed in the liquid, its frequency will shift downward slightly because of mass loading and it will also be damped to a greater degree. In most cases, the small but reproducible frequency shift is detected, but the drop in amplitude or even the loss of oscillation of a marginal oscillator can be detected instead.

The third discrete sensor (Fig. 8-8c) relies on the internal reflection of light in a prism. Complete internal reflection will occur if the difference of index of refraction between the prism and outside is sufficiently great and the angle of incidence of the beam with respect to the reflecting surface (surface of prism) is sufficiently small. The prism angle (about 45°) is chosen so that light is internally reflected in air. Light from the LED will be twice reflected into the photodiode detector. Since the index of refraction in liquids is substantially higher than that of air (or any gas), the difference in the indexes will be small when the prism is in the liquid, and therefore the light beam will be transmitted instead of reflected and the light hitting the photodetector will drop markedly. Readout circuits are discussed in Chap. 5. Fiber-optic coupling may be used.

A generally applicable continuous liquid-level sensor is the capacitance type, illustrated in Fig. 8-9. It is based on the difference in dielectric constant of the liquid and gas above. One of several geometric forms consists of a central metal rod in a metal tank. The capacitance increase C_x in the tank is proportional to the height of the liquid h. The capacitance variation is given by

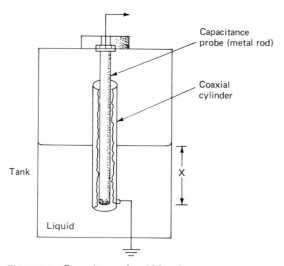

Figure 8-9 Capacitance liquid-level sensor.

$$\Delta C = \left[24 \, (\varepsilon_d - 1) \log \left(\frac{A}{B} \right) \right] X \tag{8-6}$$

where X is the liquid height (cm), ε_d is the dielectric constant of the liquid, and A and B are the outer and inner diameters, respectively. The liquid must be nonconducting or at least the resistance between electrodes must be orders of magnitude higher than the capacitive reactance at the operating frequency.

8-4 Level Measurement by Sonar

Level measurement by sonic echo time is effective in larger-scale applications. Pulses of sound waves are transmitted through the air and reflect from the liquid (or solid) surface back to the receiver (Fig. 8-10a). Alternatively, the sound wave can be transmitted from the bottom through the liquid. The echo time T_e is immediately converted to distance X by the relation

$$X = \frac{cT_e}{2} \tag{8-7}$$

where c is the velocity of sound (1460 m/s in water; 331 m/s in air).

Sonic waveforms can be complex (Fig. 8-10d), but the echo time (time between rising edges of the transmitted and received pulses) is easily discerned (within one or two cycles). It is important to ensure that the echo from the surface measured arrives first (not a spurious echo such as from the tank side). Focusing of sound waves may be required to meet this condition. Fortunately, the desired longitudinal waves travel fastest, so that the leading edge of the signal is least affected by spurious echos.

At low frequencies (3 to 30 kHz), the transmitter and receiver can be a loudspeaker and microphone but normally ultrasonic (30 to 300 kHz) piezoelectric (quartz or barium titonate) crystals are used for both. If a single crystal is used, an electronic switch changes the crystal from transmit to receive mode (Fig. 8-10c). The transmit pulse train must contain ac components only at the desired transmitting frequency to avoid excitation of spurious crystal resonances, and a tuned LC resonance circuit is usually employed for this purpose (Fig. 8-11a). Since the received signal is smaller than the transmitted signal (factors of 10^3 to 10^7), various techniques are required to damp out the crystal oscillations between the transmit and receive phases of operation. Often the receiver amplifer gain is automatically increased with time following the transmit pulse since later echos are weaker. A comparator on the receiver amplifier output detects the arrival of the echo and resets the echo time flip-flop (Fig. 8-10c) which was set by the start of the transmit

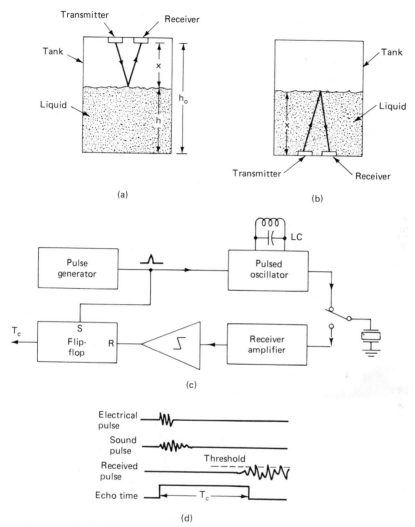

Figure 8-10 Ultrasonic (sonar) liquid-level sensor: (*a*) air path transmission; (*b*) liquid path; (*c*) block diagram of readout; (*d*) signal waveforms.

pulse. Integrated circuits containing the circuit blocks shown are made by Polaroid, Texas Instruments, and National Semiconductor.

Frequency selection depends on range (distance) and surface smoothness. Sound attenuation (transmission loss per meter) is much higher at higher frequencies. Maximum range in air is about 5 to 50 m at 30 kHz. However, higher frequencies are easier to focus (Fig. 8-11*b*, *c*) because the transducer width should be much larger than the sound wavelength for a sonic lens to be effective.

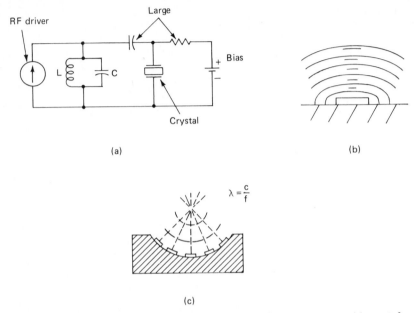

Figure 8-11   Ultrasonic driving crystal: (*a*) circuit; (*b*) wave pattern; (*c*) crystal positioning for sound focus.

Surface roughness (particle size for solids, ripples for liquids) will affect the direction of the reflected sound wave if the wavelength λ is comparable or smaller than the dimension of the surface irregularities. For long wavelengths (low frequency), the wave will reflect at the mirror angle which is the averaged surface. At short wavelengths, the reflection will be diffused. Reflected intensity for the desired echo must be high enough in the receiver direction to assure a strong signal, while early spurious reflections should be small.

8-5   Humidity Sensors

Humidity sensors measure the amount of water in a gas, especially air, in terms of mass fraction (absolute humidity) or in terms of percentage of vapor saturation (relative humidity at a given temperature). Most sensors rely on the absorption of water onto a surface. Some physical property of the surface (e.g., resistance or capacitance) is determined as a function of the moisture content or humidity of the adjacent gas phase.

Humidity sensors which depend on electrical properties are shown in Fig. 8-12. The one in Fig. 8-12*a* depends on the increase in conductance of a hygroscopic film. As the humidity increases, the absorbed water

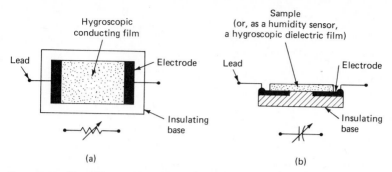

Figure 8-12   Humidity sensors: (*a*) conducting film type; (*b*) capacitance type.

increases and thus the resistance between the electrodes decreases. Older types utilize an ionic hydrate salt such as LiCl, but newer types are based on a hygroscopic polymer such as sulfonated polystyrene or a ceramic such as zirconium oxide. Generally the conductivity increases approximately exponentially with relative humidity (Fig. 8-12*c*), but the output is linearized by the parallel resistor technique (Fig. 4-6). Hysteresis (different readings for rising and falling humidity) is rather high (2 to 10 percent). Temperature coefficients are also high (5 percent per degree Celsius) and temperature compensation is necessary. A further complication is that ac excitation is required to avoid electrode polarization (a problem with all ionic conductors).

Humidity sensors based on the capacitance increase that occurs when water absorbs on gold surfaces or polymer show a reproducible response, but at this time they are costly and are not widely available.

Dew-point devices measure absolute and relative humidity indirectly but fairly accurately. A mirror is cooled (usually by a thermoelectric cooler) until fog develops (Fig. 8-13), as detected by a reflected light beam. This dew- or frost-point temperature is recorded. From the dew-point and the room (gas) temperature, the relative and absolute humidity are found by reference to standard tables.* The mirror is then heated to drive off the condensed vapor and the process is repeated. Normally, the mirror is mounted in a closed chamber through which the gas must be drawn, an inconvenience.

A standard method of determining relative humidity is the wet- and dry-bulb method, as incorporated in the sling psychometer. Wet gauze is wrapped around a thermometer bulb, and a fast flow of room temperature air is passed across it to cool it. The temperature drop

---

* Humidity tables are listed in the *Handbook of Physics and Chemistry*.

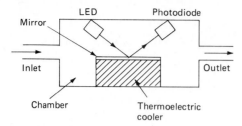

Figure 8-13 Dew point humidity sensor.

(depression, typically 2 to 20°C) is recorded along with the room temperature. The humidity is obtained by reference to tables.

8-6 Moisture Sensors

Moisture content of solids is a parameter of interest in manufacture and storage of food and other materials. One method is to insert a humidity sensor covered with a vapor-permeable jacket into the material containing moisture (e.g., grain) as indicated in Fig. 8-14a.

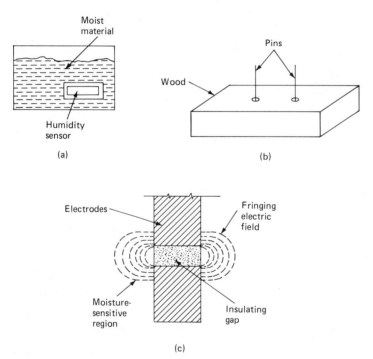

Figure 8-14 Moisture sensor in solids: (a) pore humidity method; (b) conductivity method; (c) capacitance method.

The partial pressure of a water vapor is often proportional to the concentration of water in the material (Henry's law), at least within a certain range, and therefore, the humidity sensor (Sec. 8-5) reading will be proportional to water volume fraction (moisture content by volume). The constant of proportionality differs greatly with composition.

Another method is based on the electrical properties of the mixture containing water. Both the conductivity, or conductance, and dielectric constant, or capacitance, vary with moisture. The conductance is easier to measure but is much more dependent on composition. The dielectric constant and conductivity are less material dependent at higher frequencies (>1 MHz). As an example of the conductivity method, the conductance between pins inserted into a wooden board is observed to be roughly proportional to moisture content in the normal dry-wood range (8 to 16 percent).

Capacitance between electrodes that are filled with the moist material as a dielectric increases with moisture content. Readout methods for capacitance sensors have been discussed above (Sec. 6-4). Fringing electrodes (Fig. 8-14c) allow insertion of the sensor into the moist material.

Chapter

9

Chemical and Biological Electrodes

The purpose of chemical or biological electrodes is to make good electrical contact with a solution, tissue, or part of a cell. Practically all chemical electrodes of interest involve contact with an ionic aqueous solution typified by a 0.1 N KCl or NaCl solution. Contact with biological systems likewise is via the ionic solution surrounding cells or inside them. Both chemical and biological electrodes must act as an interface between a salt solution and the metal lead-in wires from an electronic instrument. A good contact will have a low or known constant voltage across the interface, and the resistance will be as low as possible. With a number of exceptions, this is not accomplished successfully simply by sticking metal wires into a beaker or tissue. The problem usually encountered is that the voltage difference between the wire and solution is not constant unless the electrode is reversible. For some applications a dc drift in interfacial potential is tolerable, but for most cases a reversible electrode is required.

9-1   Reversible Electrodes

Reversible electrodes will be discussed here in terms of a specific example, the silver–silver chloride (Ag|AgCl) electrode pictured in Fig. 9-1. Silver metal (wire) coated with AgCl is immersed in a water solution containing chloride ions. A reversible chemical reaction occurs at the interface between the solution and metal. In the forward reaction Cl^- in solution reacts with Ag (metal), AgCl (solid) is deposited, an electron is transferred to the metal, and a corresponding

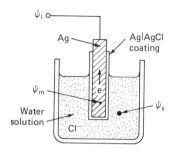
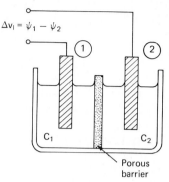

Figure 9-1 Reversible electrodes: $Ag + Cl^- \rightleftharpoons AgCl + e$: (*a*) potentials at one electrode (theoretical); (*b*) potential between a pair of electrodes.

current flows through the wire to the external surface. In the reverse direction, electrons flow from the metal, AgCl is removed, and Cl^- is transferred into the solution (Ag metal is deposited). The process can be visualized as an interchange between the charge carriers in water (Cl^-) and the charge carriers in metal (electrons) by a specific, reversible process at the interface.

As expected for any chemical reaction, the standard free energies of the charged species (Cl^- and e^-) are unequal in general, and therefore the Cl^- deposit or removal involves an energy for the reactants and products

$$\mu_{Ag} + \mu_{Cl} = \mu_{AgCl} - \mathscr{F}\psi_i \qquad (9\text{-}1a)$$

where ψ_i is the potential difference between the solution and metal and is the faraday (electron charge times Avogadro's number). Chemical potentials of solids (μ_{Ag} and μ_{AgCl}) are constant at a given temperature, but that of the solution depends on concentrations C, here that of the chloride ion. Reexpressing the above equation in terms of concentrations, we have

$$\mu_{Ag} + \mu_{Cl}^\circ + RT \ln C = \mu_{AgCl} - \mathscr{F}\psi_i \qquad (9\text{-}1b)$$

where μ_{Cl} is standard free energy* of chloride ions, a constant at a given temperature, T is absolute temperature, and R is the gas constant. After dividing by $\mathscr{F}$ and collecting the constant terms into a single constant ψ_0, we obtain an expression for the potential difference between solution and metal ψ_i as a function of chloride ion concentration C

---

* The usual approximation of expressing μ in terms of concentrations rather than activities has been made for the sake of simplicity.

$$\psi_i = -\frac{RT}{\mathscr{F}} \ln C + \psi_0 \tag{9-2}$$

This relation is not very practical in this form because the constant ψ_0, which is a kind of contact potential, is not accurately known or experimentally measurable directly, although it can be estimated theoretically. Of course since two electrodes must be present for a complete circuit, ψ_0 simply cancels in the expression for the potential difference between the two identical electrodes Δv_i. For Δv_i to be nonzero, at least one of the following conditions must be true: (1) the two electrodes are not identical (one might be Ag|AgCl and the other Hg|HgCl, perhaps); (2) the concentrations of chloride ions in which the two electrodes are immersed are not identical (a porous barrier or membrane will maintain a concentration difference); or (3) there is voltage difference in the solutions in which the electrodes are immersed caused by membrane or other diffusion potential. An expression for the potential difference Δv_i between a pair of electrodes immersed in solutions of different concentrations C_1 and C_2 and separated by a membrane with a potential across it Δv_m is

$$\Delta v_i = 2.30 \frac{RT}{\mathscr{F}} \log \frac{C_1}{C_2} + \Delta v_m + \psi_0^{(1)} - \psi_0^{(2)} \tag{9-3}$$

where $2.30 RT/\mathscr{F} = 59.12$ mV at 25°C. Usually the experimental conditions are such that only one term of this equation is nonzero. For example, if no membrane is present ($C_1 = C_2$), Δv_i represents the difference contact potential between pairs of electrodes of different metals (and metal chloride coatings). Tables listing these half-cell potentials with the hydrogen electrode as reference can be found in texts on electrochemistry.

A popular reversible electrode is the calomel electrode, which is based on the reaction between mercury (metal) and chloride ions in solution to form mercury (I) chloride (solid)

$$Hg + Cl^- \rightleftharpoons HgCl + e^-$$

The reaction is similar to that of the Ag|AgCl electrode, and the same equations apply except that the value of ψ_0 is different. Construction details for the calomel reference electrode are given below.

9-2   Reference Electrodes

An obvious drawback to the Ag|AgCl and other reversible electrodes is that they require Cl^- of a fixed concentration to be present in the

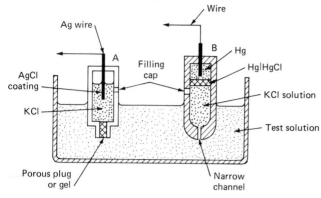

Figure 9-2 References electrodes: (*a*) Ag|AgCl type; (*b*) calomel type.

solution adjacent to the electrode. In order to make contact with solutions which contain no chloride ion or chloride ion with an unknown concentration, a reference electrode like that pictured in Fig. 9-2 is almost universally used. It consists of a reversible electrode in contact with potassium chloride (KCl) solution at a fixed concentration, which in turn is in contact with the test solution through some sort of barrier. The barrier must be such that a true liquid junction exists, but the bulk flow of the KCl solution into the test solution (due to gravity, etc.) is extremely slow. A narrow channel or porous plug (the equivalent of many channels) suffices.

At any liquid junction where bulk flow is negligible, the diffusion of ions can still take place along their concentration gradients. At the reference electrode, K^+ and Cl^- diffuse across the junction into the test solution (assuming, as is ordinarily the case, that the concentrations of K^+ and Cl^- are high inside the reservoir). When an ion difuses across a junction, it produces a junction potential. For a cation, the lower-concentration side is positive. When both cation and anion diffusion are present, their effects tend to cancel and the faster-diffusing ion determines the sign of the junction potential. By coincidence the diffusion coefficients of Cl^- and K^+ are almost equal, and therefore the potential at a KCl junction is nearly zero. For this reason KCl is almost always chosen as the salt for a reference electrode. Thus it can be seen that the potential difference between the lead-in wire and test solution is always a constant, independent of the chloride ion concentration of the test solution, in effect eliminating the concentration term of Eq. (9-3). When identical reference electrodes are used to make contact with the solution, $\psi_0^{(1)} - \psi_0^{(2)}$, Eq. (9-3) reduces simply to $\Delta v_i = \Delta v_m$.

9-3 Membrane Potentials

Practically all biological potentials are associated with membranes, especially the membranes surrounding cells. Many chemical transducers also are based on membrane voltage. Potentials develop across membranes when the concentrations of a permeable ion are unequal on either side of a membrane. A positive ion (cation) diffusing from the higher- to lower-concentration side leaves the anion (assumed incapable of penetrating the membrane) behind, and the resulting charge separation produces the membrane potential, which can be measured by a pair of reference electrodes (Fig. 9-3). The voltage difference continues to build up until at equilibrium the electric field within the membrane is strong enough to inhibit further diffusion (no external current flow is assumed). This process is similar to what occurs at (diode) pn junctions except that here cations and anions replace holes and electrons. An expression for the potential can be derived from the electrochemical potential for an ion. Since at equilibrium the chemical potentials of the two sides of the membrane are equal, $\mu_1 = \mu_2$ or

$$\mu_0 + RT \ln C_1 + n\mathscr{F}\psi_1 = \mu_0 + RT \ln C_2 + n\mathscr{F}\psi_2 \tag{9-4}$$

where C_1 and C_2 are the concentrations of the (single) permeating ion of valence n [also see Eq. (9-1)]. Solving for $v_m = \psi_1 - \psi_2$, we obtain

$$v_m = 2.30 \frac{RT}{n\mathscr{F}} \log \frac{C_1}{C_2} \tag{9-5}$$

where $N = 1$ for univalent cations, that is, K^+, and -1 for univalent anions, that is, Cl^-. This equation is known as the *Nernst equation*. Note that at 25°C, Δv_m is 59 mV for a tenfold difference in concentration and 118 mV for a hundredfold difference. If the membrane is permeable to more than one ion, Eq. (9-5) must be replaced by a more complicated expression involving the relative ion permeabilities (or diffusion coefficients or mobilities).

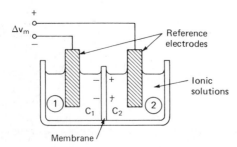

Figure 9-3 Membrane potential test chamber: sign of potential is shown for a cation.

9-4 pH Meters

Measurement and control of pH is important in numerous chemical and biological studies. pH is a measure of hydrogen ion concentration C_h defined as

$$\text{pH} = \log C_h \tag{9-6}$$

A pH meter consists of a glass electrode, reference electrode (usually calomel), and a high-input-impedance dc amplifier (Fig. 9-4). The glass electrode is made from a special glass which is permeable to hydrogen ions (only), and the thin bottom acts as a membrane. Hydrochloric acid (1 N) inside the sealed electrode is in contact with the glass membrane as well as with a reversible electrode at the top. Both the reversible electrodes inside the glass electrode and reference electrode are identical, and therefore the potential differences due to the reference electrodes are zero. The net potential across the leads connected to the amplifier is just the membrane voltage.

The potential across the membrane is given by Eq. (9-5) with $C_1 = C_h$ (test solution) and $C_2 = 1.0$ since the H^+ concentration is fixed at 1 N inside.

Substituting into Eq. (9-6), we get the pH-voltage relation

$$\Delta v_i = \frac{-2.30RT}{\mathscr{F}} \quad \text{pH} \tag{9-7}$$

As might be expected, the resistance of the glass membrane is quite high (10^6 to 10^8 Ω), and therefore a high-resistance dc amplifier or electrometer is required to make the potential measurement. A suitable amplifier is shown in Fig. 9-5. It is just a standard noninverting amplifier with an FET op-amp (for high impedance) and meter calibrated in pH units. A zero offset control to compensate for inequalities in the reference electrode contact potentials, etc., is present. Usually it is adjusted by placing the electrodes in a standard buffer solution (say pH = 7.00) and setting the meter to that pH. Often the gain is made

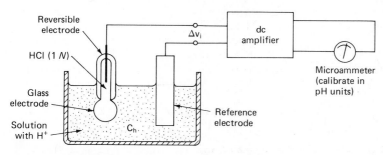

Figure 9-4 The pH meter.

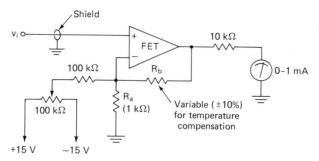

Figure 9-5   Electrometer amplifier for a pH meter.

adjustable over a narrow range through an adjustment of R_b in order to provide temperature compensation [Eq. (9-5)].

9-5   Specific Ion Electrodes

Electrodes to measure the concentration of specific ions besides H^+ have been developed. Basically they are identical to the pH meter except that the membrane utilized is permeable to specific ions to be measured. Ideally the membrane should be completely impermeable to all other ions, and the chemical engineering problem is to design such membranes.

Glass electrodes specifically permeable to Na^+ and K^+ have also been developed and are interchangeable with the pH electrodes.

Another example of a specific ion electrode is one intended to measure fluoride ion concentration. The construction of the electrode, shown in Fig. 9-6a, is based on the observation that the conductivity in lanthanum fluoride crystals involves the transport of F^-. It can be thought of as a thick membrane permeable to F^- and mounted between a test solution and a standard solution of F^- (0.01 N KF).

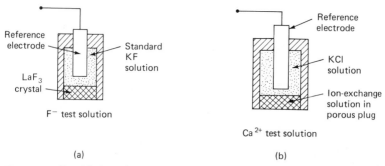

Figure 9-6   Specific ion electrodes: (a) fluoride ion; (b) calcium (ion-exchange) electrode.

When reference electrodes are in contact with both the standard test solutions, the potential developed is given by Eq. (9-5) with $n = -1$.

Calcium ion concentrations can be determined with an ion-exchanger phase separating the test solution containing Ca^{2+} and a standard solution of KCl (Fig. 9-6b). In effect the exchanger phase acts as a membrane. The liquid ion-exchanger phase consists of a porous plug filled with a standard $CaCl_2$ solution (connected to a reservoir). The common anion is Cl^-, and the cations which exchange are Ca^{2+} and K^+. For each Ca^{2+} ion which diffuses out of the exchanger phase into the test solution, two K^+ ions diffuse into this phase from the KCl region and a potential builds up across the exchanger "membrane." The potential is given by Eq. (9-5) with $n = 2$, where C_2 is the calcium concentration of the test phase and C_1 the calcium ion concentration of the exchanger phase ($CaCl_2$ concentration in the reservoir). While enjoying the advantages of other specific ion electrodes, this electrode has the disadvantages of a comparatively large size, easy chemical contamination, and Ca^{2+} leakage into the test solution.

Other ions measured by chemical electrodes include $CO_2|CO_3^-$, Br, I^-, Pb^{2+} and NO_3.

9-6 Oxygen Electrodes

Nonionic molecular oxygen (O_2) is converted to a charged product by an electrolysis process. The electrode reaction

$$\tfrac{1}{2}O_2 + H_2O + 2e \rightleftharpoons 2OH^-$$

involved a transfer of electrons from the metal electrode to the solution (Fig. 9-7a). The current flow is thus porportional to the rate of oxygen

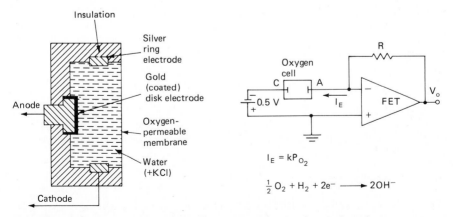

Figure 9-7 Electrolytic oxygen electrode: (a) chamber; (b) electrical readout.

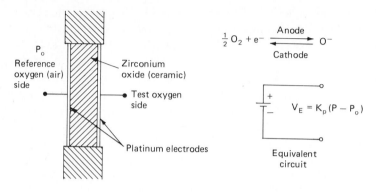

Temperature range: 550 to 1300°C
(e⁻ conductive)

Figure 9-8 Ceramic-type oxygen analyzer.

consumption and, in turn, proportional to oxygen concentration in solution.

$$I_E = k_E P_S \tag{9-8}$$

where P_S is the solution partial pressure of oxygen. Atmospheric oxygen diffuses through the oxygen-permeable (but water-impermeable) membrane to replenish the small amount of oxygen consumed so that the dissolved oxygen concentration is maintained proportionately to the partial pressure of atmospheric oxygen. By converting the current to voltage (Fig. 9-7b), the output voltage is proportional to oxygen partial pressure.

It is important to apply the proper potential (0.4 to 0.6 V) or another chemical reaction with a different stoichiometry will occur. This type of electrode is popular for biological applications. At higher temperature, conducting ceramic (ionic) sensors (Fig. 9-8) are more satisfactory. The reaction at the electrodes is

$$\frac{1}{2}O_2 + e^- \underset{\text{cathode}}{\overset{\text{anode}}{\rightleftharpoons}} O^-$$

and the voltage between the electrodes (V_E) produced is

$$V_E = K_p (P - P_o) \tag{9-9}$$

where P and P_o are the oxygen pressures on the test and reference sides and K_p is a constant.

9-7 Microelectrodes

Microelectrodes are used in biological experiments to measure voltage inside cells, especially nerve cells. The most common version is basically a glass reference electrode (disscussed above) with a hollow tip tapered to a very fine point. To make an electrode, a section of glass tube is heated until it is soft and then suddenly pulled to produce the taper. Tip diameters below 1 μm (10^{-4} cm) can be made. The inside of the tube is filled with KCl solution (3 N) by capillary action, and a reversible Ag|AgCl electrode is inserted into the large end (Fig. 9-9). In an experiment the electrode is held by a micromanipulator and pushed into the cell while the process is being watched under a microscope. A second electrode with a coarser tip is positioned outside the cell in the vicinity of the first electrode contact point. As the electrodes break through the cell wall (plasma membrane), the potential difference between the electrodes jumps from 0 to about 50 to 100 mV (resting potential, inside negative) for a nerve cell.

Usually a differential-input preamplifier (see below) is employed to measure the potential difference, and a general ground to the solution is added both to provide a ground return for the signal paths and to provide some degree of shielding. A metal-shielded enclosure is also almost indispensible.

Microelectrodes have a high internal resistance (typically 1 to 100 MΩ) due to their long taper and small tip. To avoid a signal reduction or a large dc voltage offset due to bias current, an FET input preamplifier is almost essential. Even with an FET preamplifier the high electrode resistance R_e coupled with the unavoidable shield capacitance to ground C_s results in an additional problem if rapidly changing potentials (action potentials) are to be recorded. This occurs because R_e and C_s act as a low-pass network (Chap. 10), which reduces the high-frequency components of the signal. For example, the action-potential pulse height is reduced and its width widened by a low-pass filter. To negate the effect of C_s and thus raise the frequency response, a negative-input-capacitance preamplifier (Fig. 9-10) is effective. Pos-

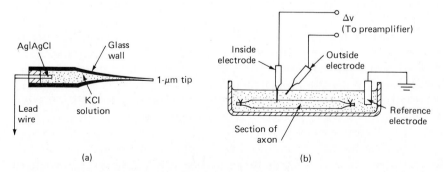

Figure 9-9 Glass microelectrodes: (a) electrode construction schematic; (b) test setup.

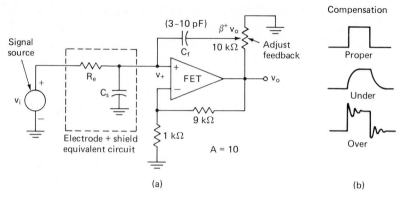

Figure 9-10 Microelectrode equivalent circuit and negative-capacitance preamplifier.

itive feedback through C_f provides the negative-input-capacitance effect.

The negative capacitance input can be analyzed by summing the currents into the node at the noninverting input v_+

$$\frac{v_i - v_+}{R_e} + C_x \frac{dv_+}{dt} - C_f \frac{d(\beta^+ v_o - v_+)}{dt} = 0 \tag{9-10}$$

where β^+ is the fraction of the voltage feedback as determined by the potentiometer setting. Recognizing that the inverting and noninverting inputs are equal and that $v_- = v_+ = v_o A$ (where A is the dc voltage gain), we see that the above equation reduces to

$$v_o = A v_i + R_e(C_x + C_f - \beta^+ C_f) \frac{dv_o}{dt} \tag{9-11}$$

The term involving dv_o/dt is the transient effect which is to be minimized. The term is zero when $C_s = \beta^+ C_f$, that is, by proper setting of the feedback potentiometer (preferably C_f is chosen so that this occurs at $\beta \approx 0.5$).

In practice the feedback is adjusted for each experiment before taking data with the microelectrode in place. A test square wave is applied to the solution, and the preamplifier output is observed on the oscilloscope. With the proper adjustment the output is also a square wave; otherwise a slow rise or ringing is observed (see Fig. 9-10b). When too much positive feedback is present (extreme overcompensation), the circuit will oscillate.

Ordinarily two microelectrodes are employed in experiments, and in this case two negative capacitance preamplifiers are desirable. Their outputs are connected to an instrumentation amplifier. Often it is more convenient to locate the reversible electrode away from the end of the glass microelectrode. In this case a salt-bridge extension is used. A thin plastic tube filled with KCl and connected to a syringe with extra solution works well. A KCl gel, which does not drip as connections are made, is preferred.

Metal microelectrodes are often preferred in applications where only an ac signal is needed, in particular where the nerve action potential and not resting potential is recorded. They consist of a thin (0.1 to 0.01 mm), stiff wire insulated except for the tip. Metals used include tungsten, platinum, platinum plus 10% rhodium, stainless steel, and silver. They are chosen for their stiffness and nontoxicity (except silver). A stiff wire is desirable because it can be guided into position more easily. Teflon-insulated wire is manufactured for this purpose. Often the tip is tapered to a point by dipping the wire tip into a solution containing the metal ion, and the metal is removed by electrolysis. Except for silver, these metals do not make reversible electrodes, and therefore the dc potential between the metal and solution or tissue is quite variable. A low-resistance contact is formed, however, and that is why they are suitable for ac applications. They are also suitable as stimulation electrodes driven by a constant-current source (usually a short pulse), since in this case the electrode contact potential is immaterial. Because metal electrodes have relatively low resistance, a negative capacitance preamplifier is not usually necessary but an FET op-amp is still desirable.

9-8   Skin Electrodes

Surface electrodes placed on the skin measure potentials originating from the electrical activity of the heart (electrocardiogram, EKG or ECG), brain (electroencephalogram, EEG), or muscle (electromyogram, EMG). The electrode must make a reasonably stable ac contact with the skin to pass these small ac signals (0.01 to 1 mV). Low resistance and low contact potentials are desirable but not essential.

A typical skin electrode, sketched in Fig. 9-11a, consists of a metal disk (or screen) held against the skin by a suction cup or strap. Usually a conducting (electrolytic) paste is smeared on the skin or electrode to provide better contact. The disk may be made of silver on which AgCl has been deposited to make a reversible electrode or the disk may be a compressed tablet of silver and silver chloride power. Stainless steel and other irreversible electrodes work well if they have a large enough contact area. Numerous variations of this type of electrode are commercially available. Electrode wires placed below the surface of the skin by hypodermic needle provide a more permanent method of attachment if an invasive technique can be tolerated.

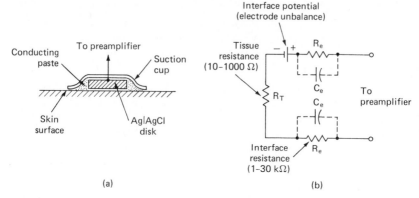

Figure 9-11 Surface or skin electrodes: (*a*) construction details; (*b*) electrical equivalent circuit for a pair of electrodes.

The contact resistance between the electrode lead wire and the tissue underlying the surface of the skin is comparatively high and variable compared with the tissue resistance. It can be modeled by the equivalent circuit of Fig. 9-11*b*. At higher frequencies (>1 kHz) the capacitive coupling between the electrode and tissue becomes important, but for practically all bioelectric signals which involve lower frequencies, the capacitive component is negligible. Actually electrode impedance is not important if the preamplifier input impedance is sufficiently high, as discussed below.

Electrode noise varies with the type of electrode material, its area, and the care with which the surface is prepared. The lowest noise is obtained with the Ag|AgCl disk. Movement of the electrodes, like that associated with muscular activity, can cause the type of noise which is in part a rapid variation in the electrode-skin contact potential. It too can be minimized by careful application of the electrode.

9-9   ECG and Related Preamplifiers

A preamplifier suitable for amplifying electrode potentials in nearly all cases must have a high input impedance since the membrane and electrode (signal-source) resistances are generally rather high. The basic instrumentation amplifier (Chap. 2) is quite satisfactory.

In ECG measurement (Fig. 9-12), a small signal (~1 mV) appearing across two electrodes is amplified by the preamplifier. A third electrode acts as a ground. As indicated previously, the ECG signal v_S, tissue resistance R_T, and electrode resistance R_e can be represented by the equivalent circuit of Fig. 9-12*b*. Unless the measurement is made in an electrically shielded room, pickup of unwanted signals, especially hum (60-Hz line) is unavoidable.

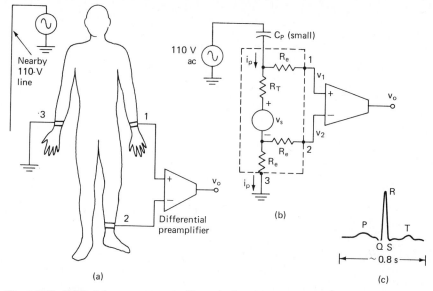

Figure 9-12 ECG: (*a*) measurement; (*b*) equivalent circuit; (*c*) signal.

The human body and the power lines (ungrounded side) act as two widely separated plates of a capacitor C_p with air as the dielectric. Due to this capacitive coupling (≈ 10 pF), a small current i_p flows into the body, through the tissue, and through electrode 3 to ground. Practically no current flows through the preamplifier electrodes 1 and 2 because the preamplifier resistance is very high. The voltage drop across tissue resistance R_T due to i_- is quite small, although the voltage drop across the grounding electrode may be substantial. In other words, the pickup voltage across leads 1 and 2 is small, assuming a third electrode ground is present to provide a path for i_p, but the pickup voltage with respect to the ground lead (1 or 2 and 3) is relatively high. The necessity of a high-impedance differential-input preamplifier which responds only to the difference voltage ($V_1 - V_2$) and not the voltage to ground (V_1) is apparent.

Frequency-response requirements differ depending on whether the purpose is diagnostic, in which case no noticeable distortion is tolerated, or monitoring, in which case all that is needed is the determination of the presence of the signal (heart rate measurements) or perhaps the detection of missing or grossly abnormal R waves. For diagnostic purposes, a frequency range of 0.02 to 400 Hz is required with a high-frequency rolloff of 6 dB per octave above 400 Hz. It is best to roll off at 12 dB per octave at low frequency. For monitoring purposes a range of 0.5 to 50 Hz or sometimes even 2 to 20 Hz is satisfactory. Because of the limited range, electrode preparation and shielding

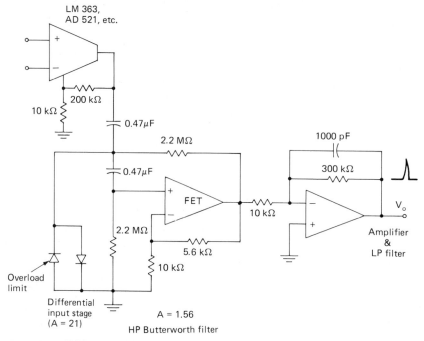

Figure 9-13   ECG preamplifier.

against hum pickup is much less critical. High- and low-pass filters are discussed in Chap. 10.

As with any amplifier with extended low-frequency response, there is a long delay before the amplifier baseline stabilizes. Large dc excursions, such as those occurring when the electrodes are first connected, are especially annoying. The amplifier of Fig. 9-13 has diode clamps to limit overload signals. Circuits which automatically zero the amplifier or provide a faster time constant for large-signal excursions are better.

Requirements for ECG, EEG, and EMG preamplifiers are rather similar. Differential input is almost essential, and only the ac component is amplified. A high-pass filter is present to block the dc electrode voltage v_J. A low-pass filter to limit the frequency response to no higher than required is also very helpful in reducing noise and other unwanted signals. Sometimes a notch filter at the line frequency (60 Hz) is added, although in the ECG signal some distortion occurs because the desired signal has frequency components in this range.

Amplifiers for the EEG must be capable of amplifying signals in the 1- to 10-μV range. Fortunately the frequency response is limited to 3 to 30 Hz, and rejection of line-frequency (60-Hz) pickup by sharp cutoff

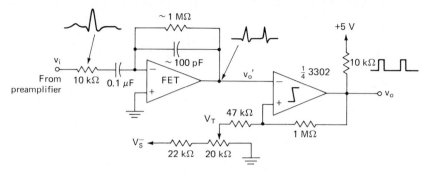

Figure 9-14   ECG-EMG signal detector.

and/or notch filters can be accomplished without difficulty. The alpha wave occurs at about 12 Hz. Gains of 10^4 to 10^5 are typical, but otherwise the design is rather similar to the monitoring-type ECG preamplifier.

EMG preamplifiers are also similar to ECG preamplifiers except that the required frequency range is 5 to 500 Hz, and because the amplitude of the signal varies considerably (0.1 to 10 mV) depending on electrode placement (muscle mass being monitored), a gain control is desirable. Sometimes only the presence or count impulses are required. For this purpose a differentiator (Chap. 10) followed by a comparator is satisfactory (Fig. 9-14). The main problems are choosing the differentiator time constants to match the signal being detected and setting the threshold V_T. Actually the differentiator or active high-pass filter is usually adjusted so that it removes most of the low-frequency components of the signal itself, with baseline removed, and the derivative seems to provide the most unambiguous input signal to the comparator v_o.

The electrode amplifiers discussed here are not recommended for use on human subjects unless ground isolation is provided on the output to eliminate the shock hazard (see next section).

9-10   Electrical Shock Hazards

Attachment of electrodes to human subjects may involve a shock hazard. Prudence and federal laws dictate that precautions be taken, particularly in the isolation of subject "ground" electrodes from equipment grounds. Approval of both experimental and commercial biomedical equipment, however innocuous, is required before it can be used on people.

Primary adverse physiological electrical effects are interference with the action of excitable membranes and nerve and muscle, and

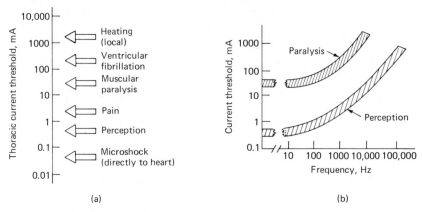

Figure 9-15 Physiological effects of a current through the thorax: (*a*) as a function of current magnitude; (*b*) as a function of frequency.

local heating. Although at the cellular level specific effects are explained in terms of membrane voltage, gross effects are expressed in terms of currents flowing through a tissue mass or even the whole body. Of course there are wide variations in response, depending on the path of current flow as well as the build and tolerance of individuals. The most life-threatening current path is through the thorax, since even moderate current levels can produce cardiac or respiratory failure. A low resistance at the skin-electrode interface implies that a particular current level can be produced by a relatively low voltage. Heating effects are important only at higher current levels (Fig. 9-15).

Most people begin to sense current flow (as a tingle) above about 1 mA. When the current through the thorax reaches 100 mA, the heart is likely to undergo ventricular fibrillation (muscular quivering rather than rhythmic beating). This potentially fatal condition often persists after the current flow stops. Still higher currents produce complete heart stoppage and paralysis of the respiratory muscles. Obviously this condition is fatal if continued, but if the current ceases, recovery is likely. In other words, if the current flow is brief, a higher current level is less likely to be fatal than a lower level. Muscular paralysis implies that the victim cannot voluntarily move away from the "hot" wire, a widely recognized problem. It is less widely recognized that these physiological effects other than heating become negligible at higher frequencies (Fig. 9-15*b*). Sinusoidal currents over 1 A at 1 MHz have been passed through a subject without sensation or ill effects. Local heating and tissue damage will occur if the current flows through a constricted area, producing a high local current density (this is the basis of electrosurgery). Unfortunately, at line frequency (60 Hz) the adverse physiological effects are comparable to those of dc. Asymmet-

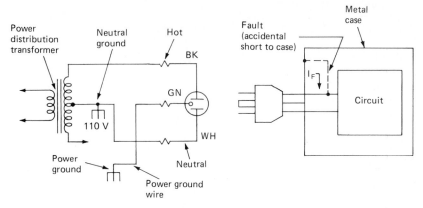

Figure 9-16   Power-line grounding and ground fault condition.

ric pulse trains, transients, or nonsinusoidal high-frequency excitations with a low-frequency component are potential hazards.

It is not immediately evident that an instrument like the ECG, which only monitors biopotentials and is not a current or voltage source to the electrodes, is any hazard at all. The problem is the ground lead or any current path to ground connected to the subject through the amplifier input leads under possible overload conditions. Should the subject accidentally come in contact with a grounded voltage source such as the hot lead of the line, current will flow through the subject and to ground through the grounded lead. The situation is best illustrated by examining normal equipment grounding techniques. Standard power-line wiring with the third-wire ground is illustrated in Fig. 9-16. Although one side of the power line (neutral) is grounded at the power distribution transformer, it is unsafe to rely on that wire for positive grounding since wiring breaks or mixups occur. The third wire runs to a separate ground, but this ground is ultimately converted back to the neutral ground. Metal instrument (or appliance) cases are attached to the ground wire. If the case were ungrounded and the hot lead accidentally came in contact with it, a person touching both case and ground would complete the circuit and receive a (possibly fatal) shock. By grounding the case, any contact of the hot wire to the case will cause a fault current to flow and the fuse to blow.

9-11   Isolation Amplifiers

The degree of isolation can be specified in terms of maximum capacitance (C_I), minimum leakage resistance R_I (and/or leakage current I_L), and minimum breakdown voltage V_{BV} between common or grounds (Fig. 9-17).

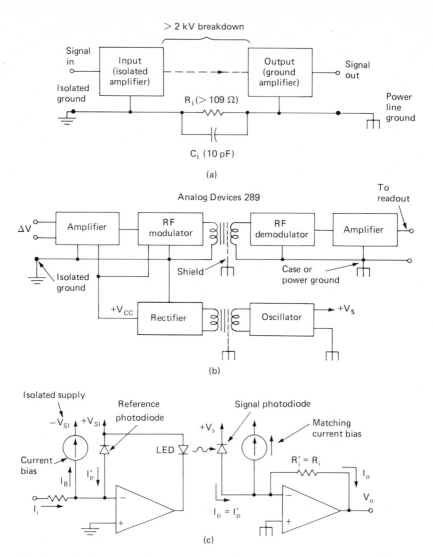

Figure 9-17 Isolation amplifier block diagrams: (*a*) isolation equivalent circuit between grounds; (*b*) transformer-coupled amplifier; and (*c*) optically coupled amplifier.

Signal coupling with the required isolation can be done by transformer or optical isolator. Special transformer construction with shielding and separation between primary and secondary windings is desirable. High-frequency (75-kHz) operation minimizes size and weight. Modulation-demodulation techniques are discussed subsequently (Chaps. 14 and 19).

Analog signal transmission through optical isolators provides excellent isolation, but basic LED-photodiode transfer characteristics limit the linearity and gain stability with temperature and time. These limitations can be overcome by providing a second, matched photodiode as a reference. Furthermore, since only one polarity signal can be transmitted, the input amplifier signal is biased by a fixed precision current I_B to allow input signals of either sign (Fig. 9-17c). A matched current is subtracted at the output amplifier to remove the bias. The two operational amplifiers force the photodiode (plus bias) currents to be equal (infinite-gain approximation) so that the currents at the input I_i and output I_o are equal. For equal resistors, R_2 R_1, the output voltage thus equals the input.

SENSOR DESIGN EXAMPLES

Example 3   An LVDT Readout

DESCRIPTION: The readout provides an output voltage proportional to the displacement of the core of an LVDT. Positive displacement produces a positive voltage, and negative displacement produces a negative output voltage since the unit incorporates a phase-sensitive detector.

SPECIFICATIONS
   Sensitivity: 10 V per millimeter of displacement (with selected LVDT)
   Range: ±1 mm
   Resolution: ±0.001 V ($±10^3$ mm), limited by short-term drift.

DESIGN CONSIDERATIONS: Assume that the LVDT has an ac differential secondary voltage of 0.18 V for a primary voltage of 6 V (400-Hz primary impedance of 100 Ω) at a displacement (from zero) of 1 mm. Excitation by a triangular-wave source (~6 V peak to peak) was chosen rather than a sine-wave source because a constant amplitude control is easier to implement with a triangular-wave source.

   With 6 V peak to peak on the primary, the amplified secondary voltage V' will be 4 V peak to peak with the amplifier of gain 8.9. The phase-sensitive detector described in Chap. 14 has a dc output v_o of +2.5 for this input. Thus the overall response corresponds to 10-V output for a 1-mm displacement from center (zero). Note that the phase-sensitive-detector output v_o reverses polarity when the displacement is in the opposite direction, an important feature of ths circuit.

   Frequency stability is not important, but the amplitude stability of source is. Variations in sensitivity from the design center are compensated by the gain or calibration control. A bias voltage on the amplifier output prevents reverse polarity on the phase-sensitive-detector capacitors.

SENSOR DESIGN PROBLEMS

10   A thermostat is needed which indicates digitally (TTL or CMOS) when the temperature is over a set point (20°C). The output should change to **1** if the temperature exceeds 25°C and change back to **0** when it drops below 23°C, that is, the hysteresis is 2°C. Use a thermistor of 10 kΩ at 25°C with a temperature coefficient of 4 percent per degree Celsius.

11   Draw a circuit which will turn on a relay (12 V, 50 mA) when a light beam

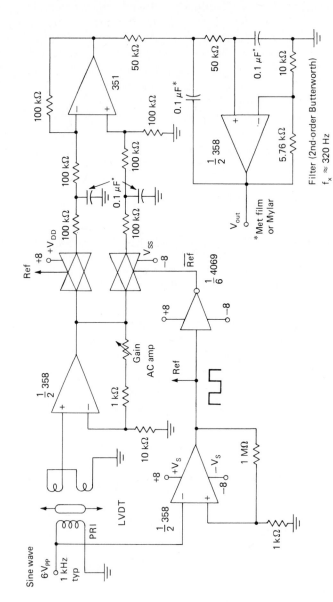

Figure E-3 LVDT Readout.

is interrupted. Assume a phototransistor is available which delivers a current of 50 μA with the light on.

12 Design an electronic readout for a pressure transducer based on a variable inductor sensor. Assume that the inductor varies from 5 mH at zero pressure to 7 mH at 1 kg/cm^2 (full-scale). The maximum recommended operating frequency is 10 kHz. A digital voltmeter (0 to 1 V dc) is available as the indicator. A phase-sensitive detector or precision rectifier as an ac-dc converter is suggested.

13 A peak-reading photometer is required to measure flash intensity for photography. Assume a phototransistor with a sensitivity of 1 μA/lm is provided. An output of 100 lm (full-scale) is desired, indicated by a microammeter-type (panel-type) voltmeter. The reading should hold following the flash until reset by a push-button switch.

14 Design a device which will read liquid level in a storage tank of gasoline using a capacitance-type gage. The tank is cylindrical (2 m in diameter, 5 m tall), and a single dipstick sensor in the center is specified. The dielectric constant of gasoline is 2.2. An accuracy of 5 percent is adequate, and a digital voltmeter of 0 to 1 V dc full-scale is available.

15 Design a battery-operated ECG monitor which will flash an LED indicator each heartbeat. Differential input is not essential with battery operation. Allow for an unpredictable electrode-polarization (dc) voltage offset up to 100 mV.

Signal Amplification and Processing

10

Active Filters

Analog signals are filtered according to frequency to a greater or lesser degree in most electronic instruments. Sometimes filtering is introduced as a result of an amplifier nonideality, e.g., limited frequency response or the necessity of blocking a dc offset. In other cases the signal itself is not in the form desired and must be shaped or otherwise processed.

This chapter is primarily concerned with active filters which affect the frequency and time response of amplifiers. Integrators, differentiators, and simple (single-pole) RC filters fall into this category, as do the sharp-cutoff-frequency filters more commonly associated with active filter design. While the analysis of simple filters is not difficult mathematically, an understanding of higher-order filters requires a good background in circuit analysis. To make it easier for the reader uninterested in mathematical details, the higher-order filters are described in terms of summary design equations.

10-1 Ideal Filters

Perhaps the most common form of signal processing is the decomposition of a signal into its frequency components by means of a filter. Ideal filter response is illustrated in Fig. 10-1. An ideal high-pass filter will pass all frequencies above a chosen break frequency f_x without attenuation but allow no frequencies below f_x to pass. A low-pass filter does the reverse, and a bandpass filter allows only frequencies within $\Delta f_0/2$ of a center frequency f_0 to pass. A notch, or band-reject, filter is the complement of a bandpass filter. Nonideal filters do not show an infinitely sharp transition at the break frequency, but real filters approaching the ideal are achievable. Actually for most purposes, a broad transition, perhaps covering a decade or more in frequency, is quite adequate.

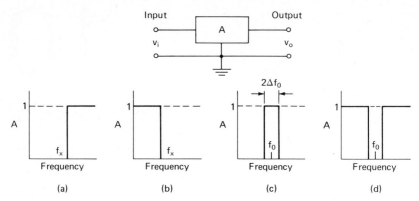

Figure 10-1 Frequency response ideal filters: (*a*) high-pass; (*b*) low-pass; (*c*) bandpass; and (*d*) notch.

Phase shifts are associated with all real filters. According to an electrical-network theorem, the slope of a passive filter gain response on a log-log scale is related to filter phase shift ϕ in radians

$$\phi = \frac{\pi}{2} \frac{d(\log|A|)}{d(\log f)}$$ (10-1)

where A is the magnitude of the response, v_o/v_i. If the frequency response of a filter (or any network) is known, the phase shift can be inferred. Phase response is important in determining the stability of control systems and other feedback networks, but since for many instrumental applications only the signal amplitude is of interest, the phase response can often be ignored.

Filter response can be expressed in three ways: in the time domain as a differential equation, in the frequency domain as a frequency response A, or in the s domain as a Laplace transform or transfer function H. In the following sections, alternate expressions are given where appropriate.

Sharp-cutoff filters can distort signals which have Fourier frequency components in the vicinity of the break frequency, an inherent drawback. An example is given in Fig. 10-2. In this respect an ideal filter does not produce an ideal result. The tailoring of filters to minimize

Figure 10-2 Response to a step function of a sharp (near-ideal) low-pass filter.

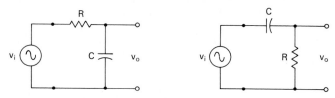

Figure 10-3 Single-section RC filters: (a) low-pass; (b) high-pass.

distortion of various classes of signals is a well-developed subject but outside the scope of this text.

10-2 Single-Stage High- and Low-Pass Circuits

Single-section RC high- and low-pass filters (Fig. 10-3), which separate the higher frequencies from the lower-frequency or dc signal components, abound in electronic instruments. Analysis of these circuits is uncomplicated if the output is unloaded. In either circuit the input current is $v_i/(R + Z_c)$, where $Z_c = -j/\omega C$ and $j = \sqrt{-1}$. The output voltage is iZ_c for the low-pass and iR for the high-pass, for which the gain ($A = v_o/v_i$) is

$$|A_L| = \frac{1}{\sqrt{1 + (\omega/\omega_x)^2}} \qquad \text{low-pass} \tag{10-2a}$$

$$|A_H| = \frac{1}{\sqrt{1 + (\omega_x/\omega)^2}} \qquad \text{high-pass} \tag{10-3a}$$

where

$$\omega_x = \frac{1}{RC} \quad \text{or} \quad f_x = \frac{1}{2\pi RC} \tag{10-4}$$

Here $|A|$ indicates the gain magnitude. If the phase shift is required, Eqs. (10-2a) and (10-3a) become, in phasor form,

$$A_L = |A_L|e^{j\phi L} \tag{10-2b}$$

$$A_H = |A_H|e^{j\phi H} \tag{10-3b}$$

where

$$\tan \phi_L = -\frac{f}{f_x} \quad \text{and} \quad \tan \phi_H = \frac{f_x}{f} \tag{10-5}$$

Plots of gain are shown in Fig. 10-4a and b for $|A|$ expressed in decibels, where $|A|$ (dB) $= 20 \log |A|$.

Interpretation of f_x as a break frequency is apparent from the

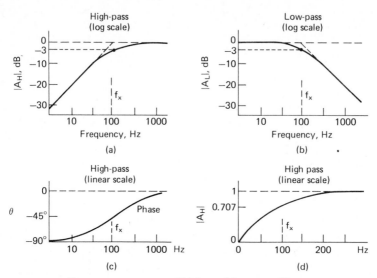

Figure 10-4 Frequency response of high- and low-pass filters.

logarithmic frequency plots but less obvious on a linear scale (Fig. 10-4c, d). At the breakpoint A_H or A_L is -3 dB, which on a linear scale is $1/\sqrt{2} = 0.707$. Note that the resistance and capacitor impedance are equal in magnitude at $f = f_x$.

A common application of the high-pass circuit is an interstage coupling of an ac amplifier, as illustrated in Fig. 10-5. It eliminates the need for compensating dc offset because the direct current is blocked between stages. All op-amps *must* have a dc return (resistor) to ground (or connection to an amplifier output which has a dc return internally) or the input will float to an unknown voltage. In this example each stage has a gain of 10, and both high-pass filters have a break frequency of 17 Hz. For the first filter it is clear that v_1 corresponds to

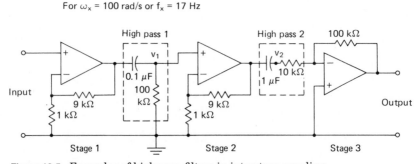

Figure 10-5 Examples of high-pass filters in interstage coupling.

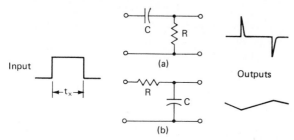

Input

Outputs

Figure 10-6 (*a*) Differentiating network; (*b*) integrating network.

v_o of Fig. 10-3, but it may not be obvious that v_2 of the second filter corresponds to the filter v_o until it is recognized that the input v_- of the third op-amp is at virtual ground.

An alternative interpretation of the high and low pass, particularly in pulse applications, is that of differentiating and integrating networks, respectively. When the filter is used as a differentiating network, the break frequency f_x is usually set so that the time constant $\tau = RC$ is shorter than the pulse width t_w, as shown in Fig. 10-6a. In this case only the edges of the pulse are passed. If τ is sufficiently small, the output signal approaches the derivative of the input signal, but this simple circuit is inferior to the differentiator (see Sec. 10-4) in this application. When it is operated as an integrating network, the time constant $\tau = RC$ is made much larger than the pulse width. The active integrator circuit (see below) is superior for this application, except at high frequencies, beyond the frequency limit of op-amps.

10-3 Integrator-Averager (Active Low-Pass)

A single-section active low-pass circuit is shown in Fig. 10-7. It is identical to the simple integrator except for the addition of the resistor

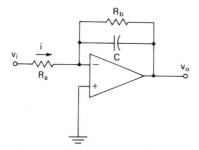

Figure 10-7 Active low-pass filter amplifier.

R_b, normally a high value, which can be thought of as a zeroing mechanism with a long time constant. It acts as an amplifier at sufficiently low frequencies or dc. It can be analyzed as follows, again under the infinite-gain approximation:

$$i = \frac{v_a}{R_a} = \frac{-v_o}{R_b} - \frac{C \, dv_o}{dt} \tag{10-6a}$$

$$I(s) = \frac{V_t(s)}{R_a} = \frac{-(1 + sCR_b)}{R_b} V_o(s) \tag{10-6b}$$

In this case the gain for a sinusoidal input, ignoring the phase shift, is

$$A = \frac{|v_o|}{|v_i|} = \frac{R_b/R_a}{\sqrt{1 + (f/f_x)^2}} \tag{10-7a}$$

which corresponds to a transfer function $H(s)$ of

$$H(s) = \frac{V_o(s)}{V_t(s)} = \frac{-1/R_a C}{s + 1/R_b C} \tag{10-7b}$$

It can be seen that the circuit acts as an inverting amplifier (Chap. 2) and low-pass filter (Sec. 10-2) in series. Its frequency response is identical to the low-pass filter response of Fig. 10-4b except that the gain axis is shifted to take into account the gain of the amplifier (R_b/R_a at dc).

The low-pass filter acts as a time averager, where the time over which the average is taken is roughly the time constant of the circuit R_bC. By contrast an integrator ideally averages over an infinite time.

10-4   Differentiator (Acting High-Pass)

The inverse equivalent of the integrator is the differentiator circuit, shown in Fig. 10-8a. Its output v_o is proportional to the derivative of the input signal v_i, which, assuming the infinite-gain approximation, can easily be derived as

$$v_o = - R_b C \frac{dv_i}{dt} \tag{10-8}$$

Note that the factor R_bC acts as a gain and the output is inverted.

If v_i is sinusoidal ($v_a \sin \omega t$), the output will be proportional to ωv_a, that is, the gain increases linearly with frequency. The implication in terms of signal quality is that high-frequency noise, ever present in the signal input, will be highly amplified and perhaps overwhelm the desired signal. To minimize noise in the practical version (Fig. 10-8b),

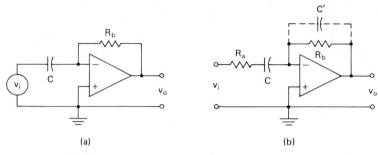

Figure 10-8 Differentiators or active high-pass filter: (a) basic circuit; (b) practical version.

a series input resistor R_a is added to limit the high-frequency gain to $-R_b/R_a$. Sometimes a small capacitor C' is also added to further reduce the high-frequency gain. Proceeding as with the active low-pass filter, we see that the gain analysis (without C') yields for a sinusoidal input.

$$|A| = \frac{|v_o|}{|v_t|} = \frac{2\pi f C R_b}{\sqrt{1 + (f/f_x)^2}} \quad \text{where } f_x = \frac{1}{2\pi R_a C} \tag{10-9}$$

In terms of a transfer function this becomes

$$H(s) = \frac{-R_b C_s}{1 + R_a C s} \tag{10-10}$$

In Fig. 10-9 the frequency response implied by Eq. (10-9), but with C' added, is plotted.

10-5 Notch and Bandpass Filters

Perhaps the easiest notch filter to adjust of those which utilize RC elements is that based on the Wein bridge (Fig. 10-10). At the notch

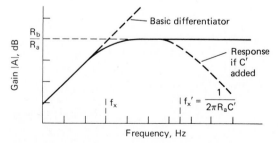

Figure 10-9 Differentiator frequency response.

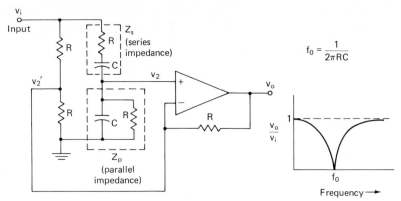

Figure 10-10 Wein-bridge notch filter.

frequency f_0, the series impedance Z_S is equal (in magnitude and phase) to twice that of the parallel branch Z_p, and thus $v_2 = v_i/3$. Resistors on the inverting side are chosen so that the inverting and noninverting gains match (at f_0), giving an output voltage v_o of zero. Close-tolerance components are desirable but not critical since minor adjustment in the notch frequency can be made by inserting a resistor in the RC branch. By varying one of the resistors on the inverting side the transmission at the notch frequency can be set very close to zero.

A twin-T notch filter (Fig. 10-11) has a notch which is sharper than the Wien-bridge filter but is considerably harder to adjust. The components must be carefully adjusted to the values shown, and the network must be driven by a low-impedance source and followed by a high-input-impedance amplifier, e.g., by unity-gain amplifiers. If these conditions are not met, the gain $|A|$ at resonance f_0 will not be zero, but with 1 percent matching of the capacitors and resistors $|A|$ will be below 0.01 at f_0.

An excellent active bandpass filter for moderate bandwidth applica-

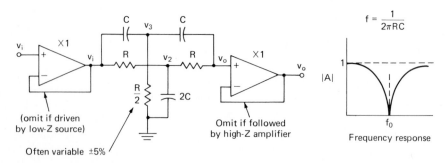

Figure 10-11 Twin-T notch filter.

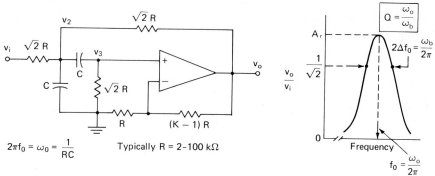

Figure 10-12 Active bandpass filter.

tion is shown in Fig. 10-12. It is a combination of the active high- and low-pass (VCVS) filters discussed in the next sections. Tuning to a specific frequency f_0 requires first the selection of the (equal-valued) capacitors and then the calculation of R (and other resistors) according to the relation given. Usually C is chosen as some convenient standard value subject to the restriction that R not be excessively large or small (loads op-amp). Close-tolerance components are required to avoid oscillation, especially for narrower bandwidths. The amplifier gain K is chosen to obtain the desired Q, that is, the reciprocal of bandwidth. At peak ($f = f_0$) the stage gain is A_R (see Table 10-1).

Active bandpass filters can be made from the switching filter (Sec. 10-9) amplifiers, or the state variable filters (Sec. 10-8). Tuned LC resonance amplifiers (Sec. 12-3) also are bandpass filters.

10-6   Butterworth Filters

The multiple-section filters designed by Butterworth are high- or low-pass filters which exhibit a sharper cutoff than simple filters. In general the sharpness of the cutoff and the approach to the ideal (Fig.

TABLE 10-1 Bandwidth 2 $\Delta f/f_o$ and Stage Gain A_R of the Active Bandpass Filter as a Function of Amplifier Gain K

K	$\dfrac{2\,\Delta f}{f_o}$	Q	A_R
(2.00)	—	0.7	(1.0)
3.00	0.71	1.4	3.0
3.50	0.35	2.8	7.0
3.70	0.21	4.7	12.3
3.80	0.14	7.1	19.0
3.85	0.12	9.4	25.7
3.90	0.07	14.1	39.0

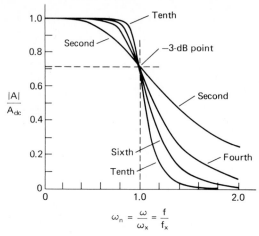

Figure 10-13 Frequency response of Butterworth low-pass filters of several orders.

10-1) increase with the number of sections. For brevity, the discussion here will be restricted to certain popular even-order Butterworth filters.

A key design requirement for Butterworth filters (and many other types of filter as well) is that the magnitude of the gain $|A|$ at the crossover (angular frequency ω_x) be -3 dB or $1/\sqrt{2}$, just as it is for a simple, single-section RC filter. Butterworth filters are composed of quadratic active filters which have this property singly or in cascade. A quadratic filter has a transfer function, or gain, expressed by

$$A = \frac{1}{1 + jb\omega/\omega_x - (\omega/\omega_x)^2} \tag{10-11}$$

If b is set equal to $\sqrt{2}$ for one quadratic filter, it can be shown that

$$|A| = \frac{1}{\sqrt{1 + (\omega/\omega_x)^4}} \tag{10-12}$$

The response of a quadratic filter, which is identical to that of a second-order Butterworth filter, is plotted in Fig. 10-13.

Two of many op-amp realizations of the quadratic filter are shown in Fig. 10-14. Both exhibit the response indicated by Eq. (10-12) except for a scaling factor or dc gain. Various combinations of capacitors and resistors will produce the same filter characteristics. While it is possible to set the dc gain A_{dc} to an even value (greater than unity) for the noninverting configuration (Fig. 10-14a), the specification of equal-valued resistors and capacitors is considered more desirable. This

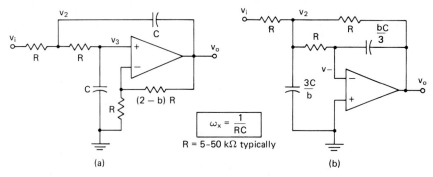

Figure 10-14 Quadratic low-pass active filters: (*a*) noninverting and (*b*) inverting configurations. Values are given for Butterworth filters ($a = 1$ and b chosen from Table 10-2).

circuit is termed the Sallen-Key or voltage-controlled voltage-source (VCVS) configuration. For the inverting configuration (Fig. 10-14*b*), the unity gain was the chosen criterion even though odd-valued capacitors are then required. Note that the parameters ω_x and b are independently adjustable. Since capacitors are available in only a limited range of values, the capacitor value is usually selected first and then R calculated from $R = 1/\omega_x C$. Close-tolerance components are generally required because the filter characteristics and stability may depend critically on several components.

Higher-order (even) Butterworth filters consist of two or more quadratic filters connected in cascade, as indicated in Fig. 10-15. While each filter is identical in form and ω_x is the same, the value of b is different for each filter. The total gain function becomes

$$A = \frac{1}{1 + jb_1\omega/\omega_x - (\omega/\omega_x)^2} \frac{1}{1 + jb_2\omega/\omega_x - (\omega/\omega_x)^2} \cdots \tag{10-13}$$

If the values of b are selected according to Table 10-2, the gain (magnitude) can be expressed as

$$|A| = \frac{1}{\sqrt{1 + (\omega/\omega_x)^{4n}}} \tag{10-14}$$

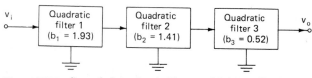

Figure 10-15 Cascaded quadratic filters as higher-order Butterworth filters. A sixth-order filter is shown.

TABLE 10-2   Butterworth Filter Constants ($a = 1$)

Order	b_1	b_2	b_3	b_4	b_5
1	1.414	—	—	—	—
4	1.845	0.7654	—	—	—
6	1.932	1.414	0.5176	—	—
8	1.962	1.663	1.111	0.3896	—
10	1.976	1.783	1.414	0.9081	0.3128

where n is the number of quadratic filters and $2n$ is the order of the Butterworth filter. Although the procedure is tedious, the value of b for any even-order Butterworth filter can be derived by substituting Eq. (10-13) into $|A|^2 = AA^*$, where A^* is the complex conjugate of A, and solving for b as the roots of a polynomial equation. However, it is not often that a filter sharper than tenth order is needed, and therefore Table 10-2 will suffice. Accurate component values are important for higher-order filters.

High-pass Butterworth filters can be designed with the same criteria except that the variable ω/ω_x is inverted, i.e., replaced by ω_x/ω in equations for A [Eq. (10-14)]. The values of b in Table 10-2 still hold. Of course the circuit configuration of a high-pass filter differs, as indicated in Fig. 10-16. The first circuit (a VCVS or Sallen-Key configuration) has a high-frequency gain A_{HF} greater than unity. The second (Fig. 10-16b) has unity high-frequency gain but has the disadvantage of low input impedance at high frequencies.

10-7   Chebyshev Filters

It must be recognized that multiple-section or higher-order RC filters have many degrees of freedom and there are many configurations and sets of component values which will satisfy the minimum design

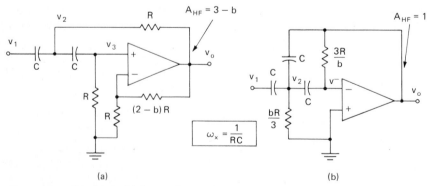

Figure 10-16 Quadratic high-pass active filters: (a) noninverting and (b) inverting configurations.

requirement of sharp cutoff. The experienced active filter designer may appreciate and take advantage of the design flexibility, but the occasional user is likely to be confused by the multiple design criteria. Here various criteria have been fixed (arbitrarily chosen) to simplify the filter design procedure even though the result may be somewhat suboptimal for a particular application. The general parameters to be chosen are the break frequency ($\omega_x = 2\pi f_x$) and the degree of sharpness of the cutoff, which is determined by the order of the filter and the choice of parameters b (and a) according to some frequency-response criteria. Once these main criteria are satisfied, there are still degrees of freedom left, and then lesser criteria are added such as equal circuit resistances, even capacitor values, or unity stage gain. Not all the lesser criteria can be satisfied simultaneously.

Another active filter, the Chebyshev filter, is similar to the Butterworth filter except that it has a sharper cutoff for a given order. It is also formed from cascading quadratic filters. The characters of the filter can be specified by the value of ω_x and b, but the choice of b's and ω_x differs. Instead of specifying different values of ω_x for each filter, it is better to define ω_x as identical for each filter and generalize the previous quadratic transfer function [Eq. (10-11)], which, written in terms of the Laplace transform, becomes

$$H(s) = \frac{H_0}{a + bs/\omega_x + (s/\omega_x)^2} \tag{10-15}$$

where a = second adjustable constant, H_0 = dc gain, and $s = j\omega$ for sinusoidal response.

It also may be expressed as:

$$A = \frac{A_{dc}}{\sqrt{1 + \varepsilon C_n{}^2}} \tag{10-16}$$

where $C_n = \cos[n \cos^{-1}(\omega/\omega_y)]$ is the Chebyshev polynomial of degree n and ε is an adjustable constant, usually less than 1.

A disadvantage of the Chebyshev filter is the appearance of gain maxima and minima, i.e., a gain ripple below the break frequency (Fig. 10-17). A Butterworth filter, by contrast, has a monotonic decrease in gain with frequency and a maximally flat response below the cutoff. The deviation from flat response is an adjustable parameter in Chebyshev filter design. It is usually expressed as gain ripple $G_r = 1 - 1/\sqrt{1 + \varepsilon^2}$. Values of a and b needed to design several Chebyshev filters are listed in Table 10-3.

Both the Butterworth and Chebyshev filters exhibit large phase shifts, especially near and above the break frequency. For applications where phase is important, a minimal phase-shift filter such as a Bessel

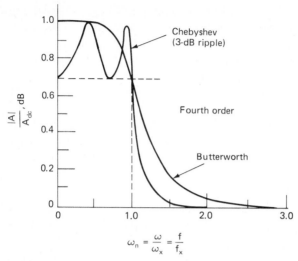

$$\omega_n = \frac{\omega}{\omega_x} = \frac{f}{f_x}$$

Figure 10-17 Comparison of a fourth-order Chebyshev filter (3-dB ripple) with a Butterworth filter.

filter may be preferred even though its frequency cutoff is not very sharp. By contrast, some filters, in particular the all-pass filter, are intended only to shift phase while allowing all signal frequencies to pass. Detailed discussions of suitable filters can be found in a number of texts on active filters.*

10-8   State Variable Filters

State variable filters are named after the mathematical technique of solving a set of linear differential equations where the derivatives ($\overset{\circ}{X}$,

---

* J. Wait, L. Huelsman, and G. Korn, "Introduction to Operational Amplifier Theory and Applications," McGraw-Hill, New York, 1975; A. B. Williams, "Electronic Filter Design Handbook," McGraw-Hill, New York, 1981; D. E. Johnson and V. Jayakumar, "Operational Amplifier Circuits: Design and Application," Prentice-Hall, Englewood Cliffs, N.J., 1982.

TABLE 10-3   Chebyshev Low-Pass-Filter Constants

Order	a_1	b_1	a_2	b_2	a_3	b_3
			1-dB Ripple			
2	1.1025	1.0977	—	—	—	—
4	0.2794	0.6737	0.9883	0.2791	—	—
6	0.1247	0.4641	0.5577	0.3397	0.9908	0.1244
			0.1-dB Ripple			
2	3.3140	2.3724	—	—	—	—
4	1.3300	0.5283	0.6229	1.2755	—	—

$\overset{\infty}{X}$, etc.) of a variable are used as a set of independent variables in a set of linear equations, usually expressed in matrix form. Any linear function, in particular the transfer function $H(s)$ of an active filter, can be formed by a sum of the set of state variables. Active filter circuits use a series of integrators (which are more noise-free than differentiators) to obtain the series of derivatives. The most popular circuit configuration consists of two integrators and a summer (Fig. 10-18). It has the following transfer function at v_{so} output (expressed in terms of the normalized variable $s_n = s/\omega_x$):

$$H(s) = \frac{H_0(1 + cs_n + es_n{}^2)}{1 + bs_n + as_n{}^2} \qquad (10\text{-}17)$$

where the numerator terms correspond to high-pass, bandpass, and low-pass responses. The output may be taken from the "low-pass" or "bandpass" outputs, rather than the summer output, in which case the numerator of $H(s)$ will contain only one term. Note that the three outputs may be considered proportional to the variable X and its two derivations $(\overset{\circ}{X}, \overset{\infty}{X})$ so that the output v_{so} is a linear combination of these state variables. Some types of filters do not require a combination of all outputs, and in this case a simplified circuit with just feedback to the inverting input (with respect to v_i) will do. For example, a second-order Butterworth filter may be implemented by making $R_7 = 1.414R$.

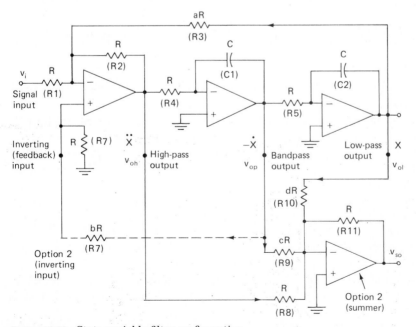

Figure 10-18   State variable filter configuration.

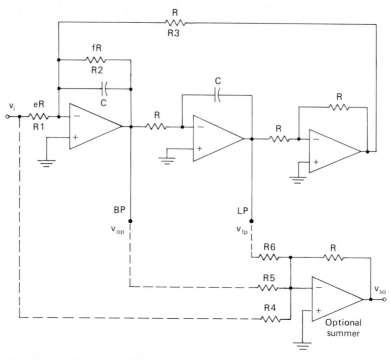

Figure 10-19 Biquadratic filter configuration.

A related filter is the biquadratic (biquad) configuration of Fig. 10-19. It has the same transfer function at the summer output (Eq. 10-17).

An example of a low-pass filter in which more than one term occurs in the numerator of Eq. (10-17) is the elliptic filter. It has zeros as well as poles in its transfer function resulting in a sharper filter than the Chebyshev or Butterworth near the break frequency (ω_x). There is, however, a gain ripple (Fig. 10-20) in the band-stop region which alternates sharp rejection and mediocre rejection. These filters are somewhat difficult to design because several parameters must be adjusted.

An active filter which is best implemented by the state variable filter is the high-Q bandpass filter. It uses the circuit of Fig. 10-19 without the summer and with the signal taken from the "bandpass" v_{op} output. In part because the designer has more freedom over component selection, the circuit can be made more stable (less sensitive to component variation and less prone to oscillation) than the VCVS circuit of Fig. 10-12. It can be shown that the resonance frequency ω_o and Q are (for $R_3 = R_4 = R$ and $R_2 = gR$):

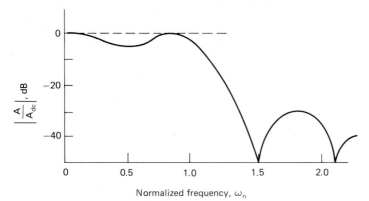

Figure 10-20 Response of an elliptic filter.

$$\omega_0 = \frac{1}{RC} \tag{10-18a}$$

$$Q = \omega_o R_2 CC = g \tag{10-18b}$$

10-9   Switched Capacitor Filters

The clock rate at which a switched capacitor filter switches determines the filter characteristics. By varying the frequency of the external clock, the break or center frequency of the filter can be changed over a moderate range. They are actually versions of the state variable filter configurations discussed above.

The basic switched capacitor integrator, shown in Fig. 10-21a, is implemented by a pair of transmission gates (Fig. 10-21b) which switch the capacitor C_s from A to B at the square-wave clock frequency f_c. When switched to the input A, the capacitor C_s is charged to voltage v_i equal to a charge of $Q = C_s v_i$. When switched to B, it is discharged to zero, since the input v_- is at virtual ground, and the charge produces a current (step) into the integrator feedback capacitor C_I. The average current I is

$$I = \frac{Q}{T_c} = f_C C_s v_i \tag{10-19}$$

If the clock frequency is high relative to the input frequency, the current fluctuation (due to the steps) can be neglected and the average and instantaneous current become equal. Like the standard integrator, $I(s) = sC_I V_o(s)$ so that the transfer function is

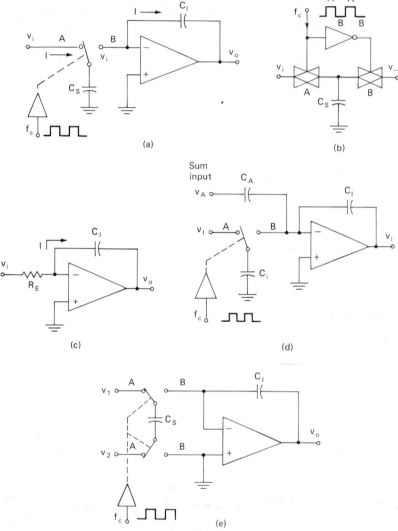

Figure 10-21 Basic switched capacitor filter circuits: (a) simple integrator; (b) implementation of switch by transmission gates; (c) equivalent circuit of simple integrator; (d) summer-integrator; and (e) differential-input integrator.

$$\frac{V_o(s)}{V_i(s)} = \frac{-f_c C_s}{sC_I} = \frac{-1}{sR_E C_I}$$

(10-20)

where $R_E = 1/f_c C_s$.

To emphasize the similarity to the standard integrator which has a resistor input, equivalent circuit and resistance R_E of Fig. 10-21c may be used. Note that the switched capacitor acts like a resistance R_E which is inversely proportional to the clock frequency. This equivalent allows the theory developed for other filters, in particular the state variable filters, to be applied to the switched capacitor filter.

Two other useful switched capacitor configurations are the combination summer-integrator (Fig. 10-21d) and the differential-input integrator (Fig. 10-21e). A voltage gain of C_A/C_I is provided at the summing input v_A (see discussion of the charge amplifier) for an ac input signal, while the other input v_I is the integrating input. The output of the differential integrator is

$$H(s) = \frac{V_2(s) - V_1(s)}{sC_IR_E} \tag{10-21}$$

A standard package of switched capacitor integrators and summers, termed a *programmable universal active filter*, has the equivalent circuit shown in Fig. 10-22. Selecting $R' = R_1$ for unity gain, the transfer function obtained is

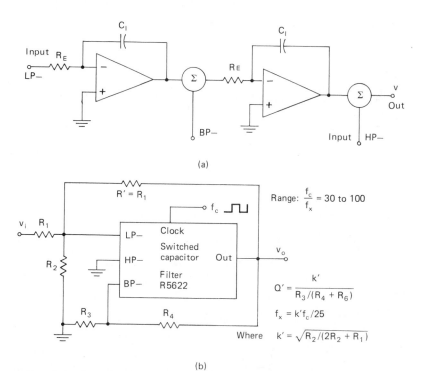

(a)

(b)

Figure 10-22 Universal switch capacitor filter: (a) equivalent circuit; (b) low-pass (Butterworth) filter configuration.

$$H(s) = \frac{1}{1 + \dfrac{1}{Q'}\left(\dfrac{s}{\omega_x}\right) + \left(\dfrac{s}{\omega_x}\right)^2}$$

(10-22)

where $Q' = 1/b$ (see also relations of Fig. 10-22). By setting $1/Q' = b = 1.414$, for example, the second-order Butterworth filter is obtained.

A related filter, the charge-coupled device (CCD) filter, will not be discussed here because it is less useful than the switched capacitor filter except for specialized applications.

10-10 Voltage-Tunable Filters

Music synthesis applications, especially, require filters (mostly low-pass) which have a wide range (100:1) voltage-controlled breakpoint. Switched capacitor filters, by contrast, are tunable only over a limited

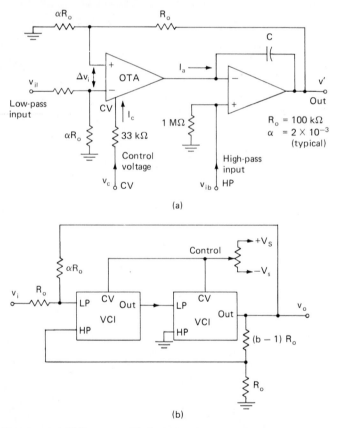

(a)

(b)

Figure 10-23 Voltage-tunable low-pass filters: (a) voltage-coupled integrator (VCI); (b) two cascaded VCIs as a state variable.

range (2:1). A common wide-range tunable filter circuit is shown in Fig. 10-23a. It makes use of the variable-gain OTA amplifier; its output is connected to integrator input to form a voltage-controlled intergrator (VCI). The OTA produces a current output I_A proportional to the input voltage difference (Δv_I) so that the integrator output v' is

$$v'(t) = \frac{-G}{C} \int \Delta v_I(t)\ dt \tag{10-23a}$$

or

$$V'(s) = \frac{-G\ \Delta V(s)}{sC} = \frac{-\Delta V_I(s)}{sCR_I} \tag{10-23b}$$

where G is the OTA gain and $R_E(v_c)$ is an equivalent resistor. Variation of R_E, and, therefore, the breakpoint ω_x over a 100:1 or 1000:1 range by control of v_c or I_c is practical. In order to change the center or break frequency by octaves, I_c is usually changed in a logarithmic manner.

Sections of the VCI are combined as a state variable filter in a manner similar to that shown for switched capacitor filters. Any type (high-, low-, and bandpass) or order of filter may be constructed by cascading sections, and all sections can be changed simultaneously, or tracked, by driving with a common control voltage.

11

Oscillators and Signal Sources

In this chapter various oscillators and waveform generators are discussed. For the most part the circuits selected are relatively simple and are intended for applications where the versatility of commercial waveform generators, in particular wide tuning range, is not required. Aside from tuning range, the most critical waveform generator characteristics are amplitude control, frequency stability, and low distortion. For most applications many of these characteristics are unimportant, and therefore the relatively uncomplicated waveform generators given here may prove quite adequate.

11-1 Wien-Bridge Oscillators

An excellent sine-wave generator at low frequencies is the Wien-bridge oscillator. Two simplified circuits are shown in Fig. 11-1. As with other sine-wave oscillators, the conditions for sustained oscillation are that (1) the phase is 0° at the operating frequency (positive feedback), and (2) the closed-loop gain (product of stage gain and feedback network gain) is exactly unity. The zero phase condition is met by the feedback network, where a phase lead of aproximately 45° by the series network (R, C) is matched by an equal phase lag by the parallel network. In order to meet the gain condition a nonlinear element is provided which reduces the gain at high signal amplitude. The stability control (amplifier gain) is adjusted so that the closed-loop gain is slightly greater than unity (divergent oscillation) for small signals. The oscillation amplitude increases until it self-limits (another type of limiter is shown in Fig. 11-1b). Some degree of distortion is inherent. Without the limiter the waveform approaches a square

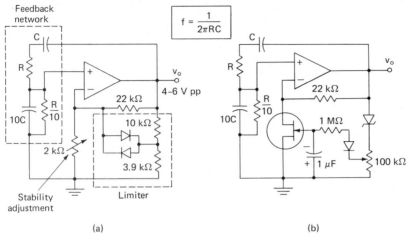

Figure 11-1   Simplified Wien-bridge oscillators.

wave, and the frequency shifts to a lower value. It can be minimized by adjusting the gain to the verge of oscillation, but at this point the stability is also minimal and the oscillation might stop with only a slight change of the feedback network with temperature or age.

In the second version of the Wien-bridge oscillator an FET type of automatic gain control is provided. The ac output is rectified, filtered, and applied to the FET gate. As the gate voltage (negative for an n-channel) rises, the FET resistance also rises and the stage gain drops, thus lowering the ac amplitude. The amplitude depends on the FET characteristics, and the regulation is modest.

More precise amplitude control and lower distortion are provided by the circuit of Fig. 11-2. The amplifier (IC1) and feedback network are the same as in the circuits above. Rectified current proportional to the signal voltage flows through D1, which is matched under constant amplitude by a current of opposite sign flowing through R_2. An unbalanced current causes the integrator output (IC2) to change and thus cause a compensating change in the FET resistance and amplifier gain. Diode D2 compensates for the voltage drop across D1, and the filter at the noninverting input of IC2 compensates for any dc bias on the output v_0.

11-2   Phase-Shift Oscillators

Phase-shift oscillators are convenient sine-wave generators at low frequency. A common version of the phase-shift oscillator is shown in Fig. 11-3. Its principle of operation is similar to that of the Wien-bridge

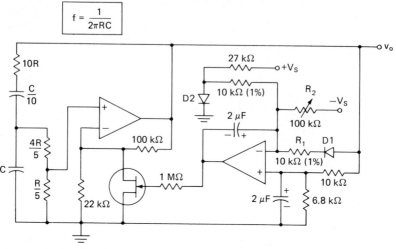

Figure 11-2 Amplitude-controlled Wien-bridge oscillator.

oscillator except that here the amplifier is inverting and the feedback (phase-shift) network shifts the phase by 180° (approximately 60° per section) to meet the 0° closed-loop phase-shift requirement. The unity closed-loop gain requirement is met by adjusting the small-signal gain to slightly greater than unity (divergent condition) and limiting (clipping) the larger-signal amplitude by a nonlinear element. The diodes conduct and therefore reduce the amplifier gain when foward-biased. The point at which limiting begins is determined by the ratio of R_1 to R_2. Alternately, the limiter of Fig. 11-1a may be used.

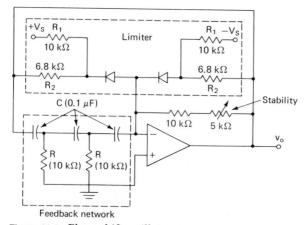

Figure 11-3 Phase-shift oscillator.

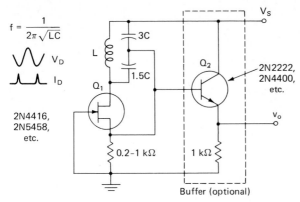

Figure 11-4   FET Colpitts oscillator.

11-3   Colpitts Sinusoidal Oscillator

At high frequencies the Colpitts oscillator (Fig. 11-4) is usually the most satisfactory. It is easily tuned by an untapped variable inductor, which at frequencies over 30 kHz is of reasonably small physical size. An emitter-follower buffer or op-amp unity-gain amplifier (at lower frequencies) may be used for output isolation. Replacement of Q1 with a BJT works well, but the base should be connected to a resistor voltage divider of 2 to 3 V dc with a bypass capacitor to provide an ac ground for the base.

The Colpitts oscillator is nonlinear and Q1 is cut off over most of the cycle. On negative (or less positive) peaks Q1 conducts, and the resonance circuit receives an impulse which compensates for resistive losses. Distortion is minimal for high-Q circuits. An IC version of the circuit, based on an amplifier described in Chap. 12, is shown in Fig. 11-5.

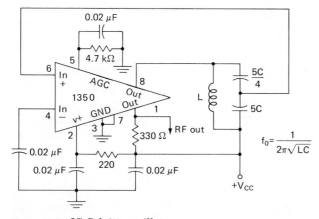

Figure 11-5   IC Colpitts oscillator.

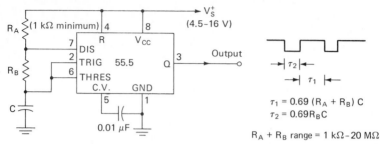

Figure 11-6 Pulse generator based on the astable IC timer.

11-4 Pulse Generators

One of the most convenient pulse generators is the astable configuration (Fig. 11-6) of the IC timer. Pulse width and frequency f_p are controlled by the indicated RC products by adjusting the ratio of R_A to R_B. For narrow pulses, $R_B \ll R_A$ and $f \approx 1.1/R_A C$. Since R_A cannot be zero, this circuit will not produce a completely symmetric square wave. Operation of IC timers is discussed in Chap. 16.

Low-cost pulse generators can be constructed out of CMOS digital devices. The second inverter of the circuit of Fig. 11-7a alternately charges and discharges the capacitor through R. When the voltage at V_t passes through the half-logic level of the first inverter (IC1), its output V_1 switches, causing the output of the second inverter V_o to change more rapidly in the same direction. Since the voltage range at V_t can be twice the supply voltage, a limiting resistor (R') is usually added so that CMOS protective input diodes do not clip the pulse. Frequency stability is relatively poor because the digital threshold levels are imprecise and temperature-dependent. Note that the relative pulse width can be adjusted if the diode circuit option is added.

A second type of CMOS oscillator is based on the hysteresis of the Schmitt trigger (Fig. 11-7b). The time required to charge (or discharge) the capacitor from one threshold level to another is half the period of oscillation. A third type of pulse generator is based on the flip-flop (Fig. 11-7c). It is well suited for short pulse generation.

Triggered pulse generators are identical in function to the monostable or one-shot multivibrators and to timer circuits discussed in Chap. 16. An op-amp type of pulse generator has been discussed previously (see Fig. 2-26). Voltage-to-frequency convertors (Chap. 15) are another type of pulse generator.

11-5 Square-Wave Generators

Several versions of a square-wave generator have been discussed. Both the op-amp version and digital IC (Figs. 11-6 and 11-7) of the pulse

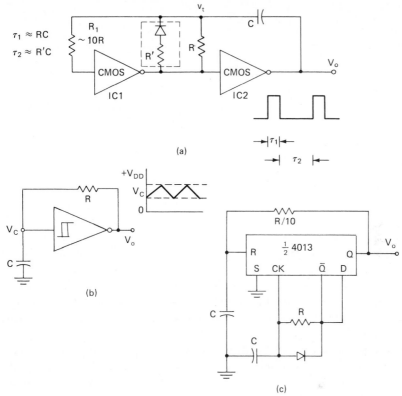

Figure 11-7 CMOS pulse generator: (*a*) two inverters; (*b*) Schmitt trigger; (*c*) flip-flop.

generators have adjustable width and can be converted into an approximately symmetric square-wave generators with careful adjustments.

Other versions of the square-wave generator are discussed as an aspect of the triangular-wave generator (Fig. 11-10) and of the IC function generator (Fig. 11-17).

If high symmetry is required, a suitable circuit consists of a frequency divider (single-stage binary counter) driven by a stable pulse generator at twice the output frequency (Fig. 11-8).

A satisfactory square-wave generator consists of a stable sine-wave

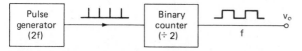

Figure 11-8 High-symmetry square-wave generator.

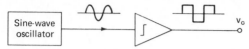

Figure 11-9   Square-wave generator.

oscillator followed by a fast (digital-style) comparator, as illustrated in Fig. 11-9. Symmetry can be trimmed by adjusting the offset voltage or by returning one input of the comparator to a variable voltage source instead of to ground.

11-6   Triangular-Wave Generators

Nearly all waveforms which involve a linear rise or ramp utilize an integrator connected to a fixed or switched voltage source. A simple ramp generator is shown in Fig. 11-10. Here IC2 is an integrator, and IC1 is a hysteresis-type comparator with a threshold determined by the ratio of R_3 to R_4. If the output of IC1 is at negative saturation, the input to the integrator will be constant and the output of IC2 will be a positive ramp. As the ramp crosses the threshold, the output of IC1 switches to positive saturation and the output of IC2 becomes a ramp with negative slope.

Note that a square-wave output as well as a triangular-wave output is available. If R_3 is decreased, the threshold will decrease, causing the amplitude to decrease and the frequency to increase. If R_1 is decreased, the frequency will increase. An asymmetric triangular wave can be produced by the diode circuit, which allows the positive and negative slopes to differ. A sawtooth wave is produced if R_2 is small.

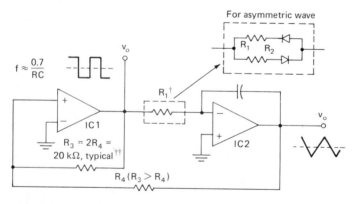

† Frequency control
†† Amplitude control

Figure 11-10   Triangular-wave generator.

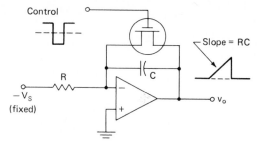

Figure 11-11   Ramp generator (step control).

For some applications, a single ramp initiated by a trigger or step input is desired. The start of the ramp in the circuit of Fig. 11-11 is controlled by the FET. It is held on by a positive voltage (usually positive saturation voltage of an op-amp) and is turned off by a negative voltage (usually negative saturation). The remaining part of the circuit is an integrator with a constant negative input, which results in a positive ramp.

Another version of the triangular-wave generator is discussed as an aspect of the function generator (see Fig. 11-19).

11-7   Crystal Oscillators

Crystal oscillators are often employed as accurate frequency or time references. Quartz crystals are commonly available between 500 kHz and 20 MHz, while the less accurate ceramic crystals cover the range from 10 kHz to 1 MHz. These piezoelectric crystals are electromechanical devices which exhibit marked electrical resonances at the mechanical resonance frequencies. When a voltage is applied, crystal constricts and, conversely, as discussed in Chap. 6, a strain produces a voltage. At the mechanical resonance, a standing wave with a half-wavelength equal to the crystal thickness occurs. Maximum mechanical oscillation amplitude occurs at or near the frequency where the electrical voltage or current is at a maximum. Because fraction and other losses are low, the Q at resonance is very high ($>10^4$) and the bandwidth correspondingly very narrow. Quartz crystals in particular are chosen as a frequency reference because they exhibit a very low change in resonance frequency with temperature (typically 0.01 percent per degree Celsius).

The electrical equivalent circuit at a crystal near resonance is shown in Fig. 11-12. Mass is modeled by an inductance L_m, elasticity by C_m, and loss by R_m for the mechanical effects. Crystal capacitance C_x is effectively in parallel. Both series (f_{os}) and parallel (f_{op}) resonances occur, as described in the reactance graph (Fig. 11-12b). These are

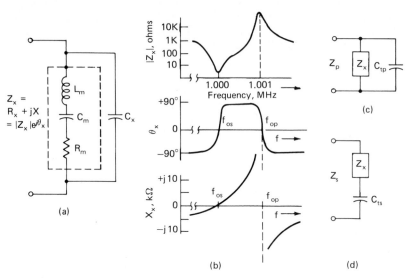

Figure 11-12 Crystal impedance: (*a*) equivalent circuit; (*b*) impedance near series and parallel resonance; (*c*) parallel resonance; and (*d*) series resonance.

closely spaced in frequency (perhaps only 0.1 percent apart), but should be distinguished in high-accuracy specifications. Note that the reactance is inductive (positive) near resonance (between f_{os} and f_{op}). By adding an external tuning capacitor (negative reactance) in parallel or series (Fig. 11-12c, d), the composite circuit resonance frequency can be shifted slightly (resonance occurs where the net reactance is zero for series resonance and infinite for parallel resonance). Since the effective inductance (or reactance) covers a fairly wide range near resonance, the value of the external capacitance is not critical. Oscillation occurs where the composite circuit is at resonance ($|A_L| > 1$ at zero phase shift). The frequency will shift slightly, and the reactance will change, until this condition occurs. Thus it is not difficult to obtain oscillation with many types of crystal oscillator configurations.

A standard series resonance oscillator circuit is shown in Fig. 11-13. It employs quasilinear amplifiers, as discussed in Chap. 12. The two inverters produce a gain of 30 to 1000 and a phase shift near zero. Oscillation occurs on the negative (capacitive) reactance side of series resonance so that the phase shift across the feedback network (Fig. 11-13b) is near zero. Thus the conditions for oscillation are present. Often a buffer amplifier is added to square up the output waveform.

A popular configuration is the Pierce oscillator (Fig. 11-14) because it requires only one inverter. It operates with the crystal in the positive inductive reactance region near parallel resonance (actually between series and parallel resonance). The π feedback network provides a

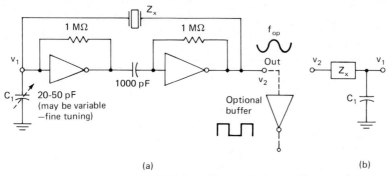

Figure 11-13 Series resonance CMOS oscillator: (*a*) circuit; (*b*) equivalent.

$-180°$ phase shift which, combined with the inverter $+180°$, produces the necessary net loop phase shift of zero. Most of the $-180°$ shift in the π network is provided between v_2 and v_1, a consequence of the ratio of positive crystal reactance to negative capacitor (C_1) reactance. The balance is contributed by the R_2C_2 part $(V_o$ to $v_2)$. Usually R_2 is omitted (replaced by a short) in the actual circuit since the internal resistance of the CMOS driver is sufficiently high.

A TTL version of the series oscillator is shown in Fig. 11-15. The parallel (single-inverter) version is possible but more difficult to make

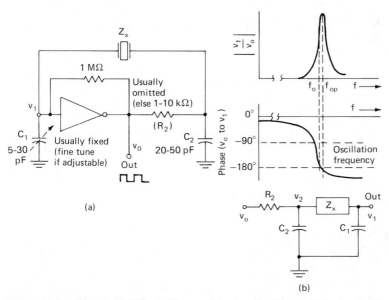

Figure 11-14 Pierce CMOS oscillator: (*a*) circuit; (*b*) equivalent circuit near resonance.

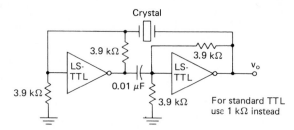

Figure 11-15   Series resonance TTL crystal oscillator.

operational since the loading by the bias resistors, as well as crystal impedance, has to be taken into account simultaneously.

The transistor Pierce oscillator (Fig. 11-16) is a convenient circuit which works under a wide variety of conditions. It is functionally identical to the circuit of Fig. 11-14. The similarity also to the Colpitts oscillator (Fig. 11-4) becomes apparent once crystal impedance (near resonance) is replaced by an inductor. With proper transistors, operation at very high frequencies is possible.

11-8   DC Signal Sources

There is often a need for a regulated or variable dc signal source in addition to the usual regulated power supplies (Chap. 21). Two semiprecision circuits are given in Fig. 11-17. The first is simply a variable potentiometer across the regulated supply followed by a unity-gain amplifier to eliminate loading effects. Of course, the output regulation is no better than that of the main supply $(+V_S, -V_S)$.

Improved stability can be achieved with a voltage-reference Zener diode. Most high-stability diodes regulate between 6 and 9 V, a region where the temperature coefficient is small. Best regulation is achieved if the load across the diode is constant and the main supply is also regulated. Higher or lower voltages than the reference are achieved by

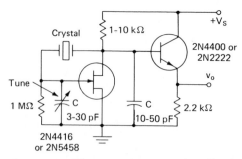

Figure 11-16   Pierce transistor crystal oscillator.

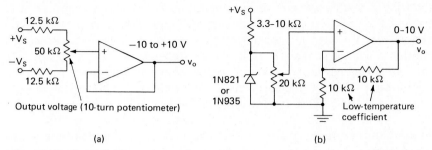

(a)   (b)

Figure 11-17 Controlled dc sources (semiprecision): (*a*) supply-reference; (*b*) Zener diode.

potentiometer voltage dividers followed by an amplifier (perhaps only unity-gain).

Precision voltage reference sources are the best method of obtaining a stable voltage source (Fig. 11-18). They have excellent temperature stability (0.01 to 0.001 percent per degree Celsius) and long-term

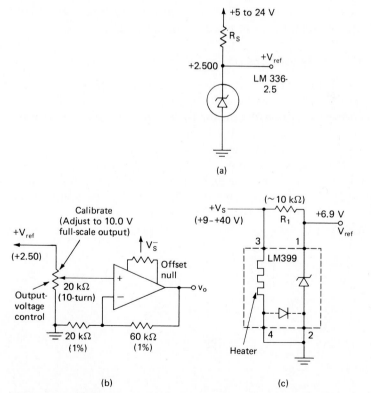

Figure 11-18 Precision voltage reference sources: (*a*) basic regulator; (*b*) variable; and (*c*) temperature-stabilized.

stability [3 to 20 parts per million (ppm)]. Most have two terminals and are used as a normal Zener diode, although they are ICs of some complexity, so that the supply voltage regulation is less critical (still a regulated source is desirable). Dynamic impedance R_Z is very low (<1 Ω). The series resistor R_s is adjusted so that the current through the regulator is within the desired range (0.4 to 2 mA). The initial voltage tolerance of the LM336 is small (0.5 percent), but for other types it can be much larger (up to 15 percent). Some means of calibration or adjustment of the output to a standard voltage is often necessary. In the circuit of Fig. 11-18b, a variable series resistor (trim-pot) allows adjustment of the voltage across the 10-turn potentiometer to 5.0 V and thus the circuit output v_o to +10.000 V full-scale. Previous nulling of the amplifier offset voltage to within ±1 mV is necessary. Since 10-turn potentiometers are usually linear to within ±0.1 percent, the output voltage should be accurate to ±10 mV or better.

Higher-precision regulators require temperature control of the internal Zener diode. Fortunately IC regulators with built-in heaters are available, as indicated in Fig. 11-18c. The heater section supply can be connected to any 9- to 40-V supply of adequate current capacity (200 mA peak). Long-term stability of about 10 ppm is achievable.

11-10   IC Function Generators

Because the need to generate a triangular or square wave or related waveform occurs often, a specialized IC has been designed to meet this need. The circuit is functionally similar to the triangular-wave generator of Fig. 11-10 but is contained within one package. Typical external connections are shown in Fig. 11-19. Features of this VCO circuit are simplicity, accurate linear outputs, high stability, and ease of frequency adjustment. A further feature is the ability to frequency-modulate the output frequency with high linearity by the application of a modulation voltage to the threshold control.

Voltage-controlled IC sine-wave generators currently available pro-

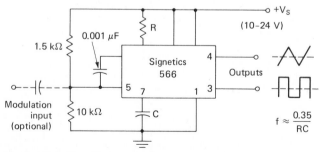

Figure 11-19 Function generator.

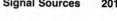

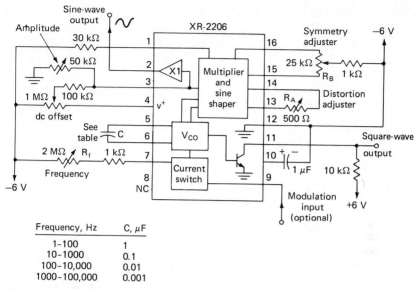

Figure 11-20 IC sine-wave generator.

Frequency, Hz	C, μF
1–100	1
10–1000	0.1
100–10,000	0.01
1000–100,000	0.001

duce a waveform of fair quality over a moderately wide frequency range. The waveform generator of Fig. 11-20, for example, produces a reasonably undistorted output over a frequency range of 100:1 by adjustment of R_F. These devices synthesize or approximate the sine wave by a multiplier, a circuit which is frequency-insensitive. Some filtering, however, may be necessary to smooth the waveform. Resistor trimming (R_A, R_B) reduces distortion to under 1 percent.

Stepwise or digital synthesis of a sine wave is possible with the

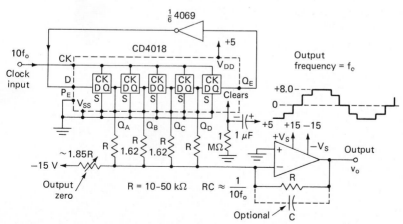

Figure 11-21 Digital sine-wave generator.

circuit of Fig. 11-21. It is basically a five-section serial register connected as a ring or Johnson counter. Except for the initial cycle (following clear on power up), each stage has the same pattern; 1 for 5 clock cycles followed by 0 for 5 cycles. The pattern is delayed by one clock cycle between stages. Zero crossing on the sine wave corresponds to turn-on of only the R and $1.62R$ resistors (in addition to the output zero resistor) into the summing amplifier. Expansion to N stages with a correspondingly better approximation to a sine wave is easily achieved by adding more stages. Resistors for each stage are given by

$$R_N = \frac{R \sin\left[\pi(N-1)/N\right]}{\sin\left(\pi/N\right)} \tag{11-1}$$

The lowest harmonic is $(2N-1)\,f_o$ with an amplitude signal equal to $1/(2N-1)$ times the fundamental.

Sine waves or, in fact any arbitrary function, can be generated by means of data stored in digital memory (EPROM or RAM) which is converted to an analog signal by a D/A converter. The address lines are incremented at a fixed rate to provide the desired sweep rate or period. This circuit is discussed as a design example (E-6).

High-Frequency Amplifiers

For most purposes the division between the low- and high-frequency operating regions of ICs is within an order of magnitude of 1 MHz. Circuit configurations which work well at low frequencies may fail if the signal frequency is high. Often the cause is the limited frequency response of the device, a problem which can usually be cured by selection of a higher-frequency type. The main drawback of high-frequency devices, aside from higher cost, is the necessity of careful component placement, wiring, and ac bypassing. The frustration of eliminating high-frequency oscillations with no apparent cause or trying to achieve the advertised performance can dampen enthusiasm for high-frequency devices. It is wise to select a device with a frequency response only moderately beyond that needed to accomplish the required task.

Design of amplifiers at frequencies above 10 to 100 MHz requires careful design with due consideration to wiring (parasitic) capacitances and inductances as well as the transfer characteristics of the semiconductor devices. Detailed circuit analysis, preferably by computer, is desirable, and the advice of experienced RF circuit designers is valuable.

12-1 Higher-Frequency Op-Amps

The most popular internally compensated op-amps have gain-bandwidth products of 1 to 10 MHz and slew rates of about 0.5 to 5 V/μs. For most applications a maximum operating frequency of 20 to 500 kHz results from these limitations. High-speed devices are available which

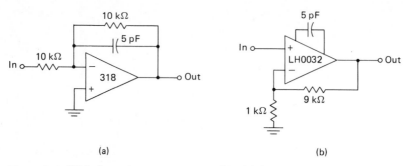

Figure 12-1 High-frequency op-amp examples: (*a*) fast inverter; (*b*) gain 10 dc 5-MHz buffer amplifier.

have the frequency limitations higher by a factor of 10 or more (Table 12-1). Some types require an external-frequency-compensation-network capacitor tailored to the particular configuration, gain, and IC type in order to obtain optimum performance (Fig. 12-1). Compensation techniques are discussed in Chap. 3.

Care must be taken in the layout and mounting of any high-frequency circuit. Component leads must be kept short to reduce inductance and stray capacitance coupling. All components, including ICs, are often soldered directly to the circuit board to reduce lead length. Printed circuit board construction is preferred, with unused areas of the board made into a ground plane (indicated in Fig. 12-2). Small surface-mounted components can be used to reduce lead length and board size. At the highest frequencies, hybrid circuit (printed resistor) construction is best. Power-supply leads are bypassed (ac-grounded) by a ceramic or other low-series-inductance capacitor close to the IC itself. In critical applications where good isolation between stages is needed, mounting the IC and associated components on a small separate circuit board inside a metal box as a shield is suggested. Input and output leads are coaxial cable.

TABLE 12-1.   High-Frequency Amplifier Characteristics

No.	Type	Gain bandwidth or maximum frequency, MHz	Input R, kΩ	Slew rate, V/μs
NE5539	Op-amp	110	710	600
AOP1510	Op-amp	150	710	500
CLC220	Op-amp	240	710	7000
CLC104	Video	1100	50	—
NE5205	Video	450	50	—
1350	RF/IF	30	10	—

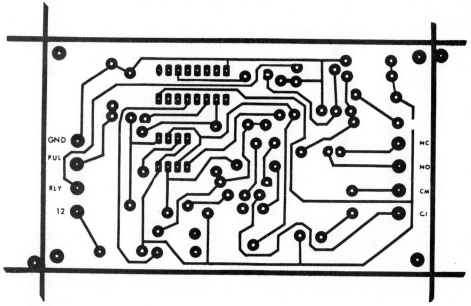

Figure 12-2 Printed circuit layout.

12-2   Wideband (Video) Amplifiers

In many applications a fixed-gain wideband amplifier is more satisfactory than the op-amp at higher frequencies. These amplifiers usually have a differential input (like op-amps), but they still cannot utilize most standard op-amp configurations because the output has a dc offset. Various types are available, one of which is shown in Fig. 12-3. Frequency response (-3 dB) at the high end exceeds 80 MHz with the feedback resistor (R_f) shown. Extension of low-frequency response is achieved by increasing C_1.

High-frequency signal transmission over an appreciable distance (over 10 to 100 cm) should be done over coaxial line. Especially at longer distances, it is necessary to terminate one end, and preferably both ends, in the cable characteristic impedance (50 Ω usually, possibly 75 or 125 Ω). Unless the signal amplitude is low (<200 mV), appreciable current will flow into the load and the cable driving amplifier must be capable of delivering this current without overload.

12-3   Tuned RF Amplifiers

Often high-frequency amplification over a narrow band is desired or acceptable. In a radio receiver, for example, the RF and IF sections are sharply tuned to pass only the desired signals. The tuned circuit is an

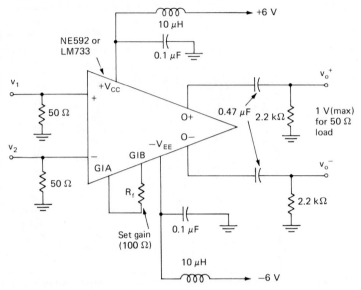

Figure 12-3 A wideband (15 Hz to 80 MHz) amplifier.

LC network, chosen to resonate at the desired frequency. Generally speaking, tuned amplifiers are more trouble-free than wideband amplifiers, in part because the capacitance to ground is tuned out by the inductance. A typical RF amplifier is shown in Fig. 12-4. Compared with that of most amplifiers, the input impedance is quite low (2 kΩ at low frequencies, dropping to about 0.3 kΩ at 60 MHz). Actually in most RF applications a lower-impedance input is often desired to match the input coaxial cable. This is easily accomplished by adding a resistor (e.g., 50 Ω) to ground near the IC input.

The behavior of a parallel *LC* resonance circuit is shown in Fig. 12-5.

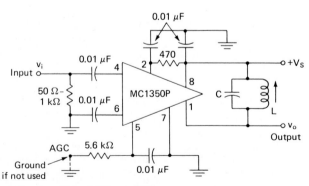

Figure 12-4 Tuned RF amplifier.

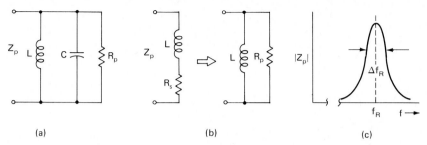

Figure 12-5 Tuned circuits: (*a*) parallel *LC* resonance; (*b*) series-to-parallel resistance equivalent transformation; (*c*) impedance as a function of frequency.

At resonance, the positive and negative reactances of the inductor and capacitor, respectively, cancel, leaving only the parallel equivalent resistance (more precisely, the imaginary part of the inverse impedance becomes zero so that $Z_p = R_p$). Although the inductor series resistance R_S is the primary cause of loss, it is usually expressed as an equivalent parallel resistance R_p (Fig. 12-5*b*). The parallel resonance frequency f_R is

$$f_R = \frac{f_o}{1 - 4/Q^2} \tag{12-1a}$$

where the no-loss resonance frequency $\omega_o = 2\pi f_o$ is

$$\omega_o = \frac{1}{\sqrt{LC}} \tag{12-1b}$$

and the circuit quality factor Q is

$$Q = \frac{R_p}{\omega_o L} \tag{12-2}$$

Note that $f_R \approx f_o$ when the Q is high and circuit loss is low.

Bandwidth, defined as the full width at half power (-3 dB) ($\Delta f_R/f_R$) is inversely related to the Q (Fig. 12-5*c*)

$$\frac{\Delta f_R}{f_R} \approx \frac{1}{Q} \tag{12-3}$$

On either side of resonance, the parallel resonance impedance Z_p drops from R_p to a much lower value. Current from the IC output (transistor collector) therefore produces maximum voltage across the *LC* network at resonance.

Sometimes transformers are used to change the output impedance

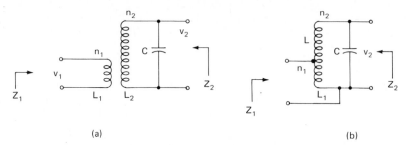

Figure 12-6 Tuned transformer (*a*) separate windings; (*b*) autotransformer.

level as well as to act as a tuned circuit (Fig. 12-6). It can be shown from circuit analysis that if the magnetic (mutual) coupling exceeds a critical value, the primary and secondary cannot be tuned separately but depend on a combination of L_1 and L_2 (as well as C). The ratio impedance (Z_1/Z_2) is obtained from

$$Z_1 = \left(\frac{n_1}{n_2}\right)^2 Z_2 \tag{12-4}$$

where n_1 and n_2 are the turns ratio. Transformers can step down the impedance, for example, to drive a low-impedance (50-Ω) line, or step up the impedance, for example, to drive a crystal at resonance.

12-4 Limiting IF Amplifiers

Limiting amplifiers provide a constant-amplitude output provided the input amplitude is above a certain minimum or threshold. They are used where the frequency, and not amplitude, of a signal is of interest, as in an FM receiver or frequency-counter applications. Basically they are simply high-gain (60-dB) amplifiers operated so that the output is saturated or clipped. In other respects they are similar to the tuned or video amplifiers discussed above. An example of such an amplifier is given in Fig. 12-7.

12-5 Inside the Open-Collector RF-IF Amplifier

Some understanding of the IC arrangement (Fig. 12-8) is helpful, especially in connection with the differential inputs and open-collector outputs. A pair of input transistors (Q1, Q2) is operated as a differential amplifier. Constant-bias voltage is provided by an internal circuit (with an external bypass capacitor) so that the inputs have a dc potential several volts above ground. Neither input can have a dc ground return,

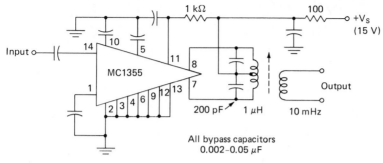

Figure 12-7 Limiting IF amplifier.

although an ac ground at one input through a bypass capacitor is often used. Outputs are open-collector and require an external load, generally a tuned LC circuit. Note that a dc return to a power supply is required for both collector outputs (Q3, Q4). A tuned transformer is often employed. If a center-tapped primary is used, the center tap is connected to the power supply and the remaining two terminals to the two amplifier outputs in a push-pull arrangement (see Fig. 12-7). A resistive load is permissible if a tuned output is not wanted. When only one output is loaded, the other output (collector) should be connected directly to the supply.

12-6 CMOS and TTL Inverters as Quasilinear Amplifiers

Although intended as a digital device, digital inverters are quite satisfactory (and inexpensive) ac amplifiers for some applications.

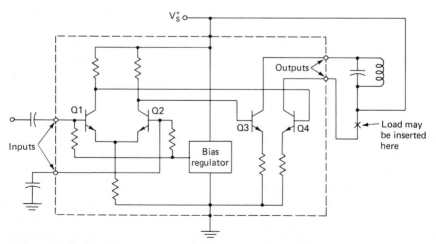

Figure 12-8 Simplified circuit for a differential-input open-collector IC amplifier.

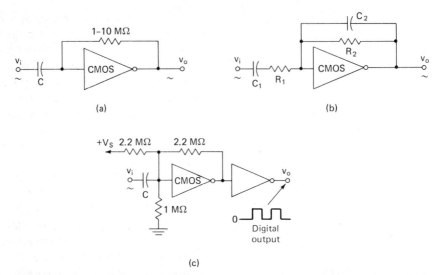

(a)

(b)

(c)

Figure 12-9 Linearly biased CMOS ac amplifiers: (*a*) basic amplifier; (*b*) controlled-frequency-response amplifier; (*c*) digital level output stage.

Biasing of CMOS devices (Fig. 12-9) consists only of a resistor to feed back the dc output to the input. Resistor value is not critical since very little dc current flows through it. The dc level of the inverter input is equal to that of the output although the ac voltages differ. With this dc feedback configuration the input and output dc potential is approximately half the supply voltage, which is intermediate between the **1** and **0** voltage levels. The exact voltage varies slightly with the particular unit and can be calculated (graphically) if the input-output characteristics are known. Appreciable supply current can be drawn under these conditions, and operation with a supply above 10 V could damage the device. The low-frequency response is determined by the input time constant RC. Actually standard CMOS inverters do not have especially good high-frequency gain. The midfrequency gain of a single stage of a standard device (UB series) is about 20, and the high-frequency breakpoint is about 100 kHz. Double-buffered devices (B series) have an extra internal stage of amplification and have a better ac gain but are less stable and may oscillate at a high frequency. Gates such as NOR and NAND are equally satisfactory, although, of course, devices with Schmitt-trigger inputs will not work. Distortion is rather high, especially at higher signal levels, and therefore this circuit should not be used where this is a consideration. High input impedance is one of the advantages of this circuit, but to avoid high-frequency oscillation, the unbuffered (UB) CMOS inverter is needed.

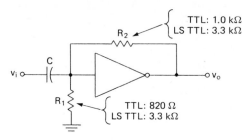

Figure 12-10 Linearly biased TTL ac amplifiers.

Several variations of the quasilinear amplifiers are shown in Fig. 12-9. Control of the high- and low-frequency cutoff points as well as midfrequency gain are accomplished by the circuit of Fig. 12-9b.

Often the objective of the linearly-biased digital amplifiers is to bring the signal level to logic level, as, for example, in frequency-counter applications. While it is possible simply to overdrive the linear stage, the output stage of Fig. 12-9c will provide a cleaner output. The input of the first stage is biased slightly away from the linear region, so that the output v_o is at **0** rather than an intermediate logic level when no ac signal is present. The last stage (optionally a Schmitt trigger) shapes the signal and acts as a buffer-driver. A fairly high level ac input signal (peak-to-peak amplitude of at least 30 percent of the supply voltage) is required, such as supplied by a previous quasi-linear stage.

Linear-mode TTL amplifier (Fig. 12-10) design is much more restrictive than CMOS amplifier design because the dc resistance to ground must be chosen to develop the required bias, taking into account the current flow out of the input terminal (typically 0.25 mA for LS-TTL and 1 mA for TTL). Some dc feedback R_2 is needed to correct for variation in the operating point of individual units (alternately R_1 can be made variable). The main advantage of TTL-type quasilinear amplifiers is good high-frequency response. A drawback is the low input impedance.

13

Analog-to-Digital Conversion

Considered here is the conversion of an analog signal or voltage which can vary continuously within a given range into a digital signal registered by a set of flip-flops or latch. In general, the cost and complexity of the analog-to-digital (A/D) converter increases with the speed and accuracy required. Conversion speed is often important in computer applications (such as the interface between an instrument and a minicomputer control) but may be of little concern in other applications, such as the digital voltmeter. Emphasis here will be on simpler types of converters. Many of the circuits are described in terms of block diagrams because detailed circuits for displays and counters are described elsewhere.

13-1 D/A Specifications

A digital-to-analog D/A converter provides an analog output v_A in proportion to a digital number on the input, usually presented in parallel binary form. Output changes when the input number changes. A/D converters do the reverse except that they require some time to convert following an input command (start) pulse. Basic converter specifications can be illustrated by the 3-bit D/A converter in Fig. 13-1. Most converters are 8-, 12-, or even 16-bit, with correspondingly higher resolution. For illustration purposes, a periodically incremented 3-bit counter is the input, and the output is therefore a staircase waveform. Ideally each step is the size of the least-significant bit (LSB). In practice errors greater than ± 1 LSB are unacceptable since the analog output voltage change might have the wrong sign (Fig. 13-1c) which in control-feedback applications is likely to result in instability. Other

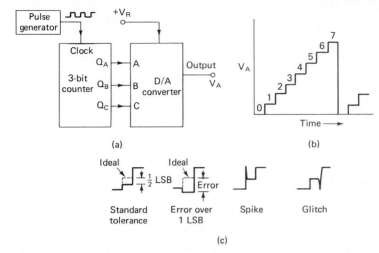

Figure 13-1   Block diagram of a D/A converter. The input to the converter is a periodically incremented by an up counter.

deviations from clean steps are spikes and glitches, most troublesome when the steps are small and are changed rapidly. A spike is usually due to stray capacitance feedthrough from the changing digital input. A glitch is usually due to a delay between switching of larger and smaller bits. As an example, consider the change from **0111** to **1000**, a net step of 1 LSB, which requires the most-significant bit (MSB) to switch. A glitch occurs if the lower bits turn off before the higher bit turns on. A low-pass filter or sample-and-hold circuit is the remedy.

The speed of D/A conversion is expressed in terms of settling time, which is usually defined as the time required for a large step to change in voltage or settle to within $\frac{1}{2}$ LSB of the final value. Higher-resolution units are slower, with typical values in the range of 0.05 to 5 μs.

A reference voltage and op-amp are necessary for all D/A and A/D converters but are often internal. Many units have dual-polarity outputs, for example, -5 to $+5$ V. Binary codes for polarity are discussed in later sections.

The speed of an A/D converter is expressed as a conversion time, which is the time between the start command and time finished. A busy or done output is always provided. Typical speeds are 1 to 100 μs. Cost rises rapidly with speed. Most A/D converters provide serial as well as parallel output.

13-2   Voltage-to-Frequency Converters

A voltage-to-frequency converter (VFC) has an output frequency which is proportional to the input voltage. Output amplitude or

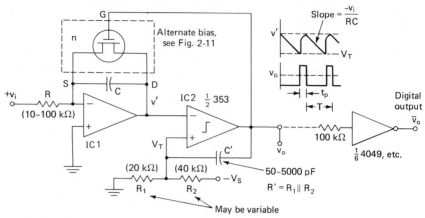

Figure 13-2   Voltage-to-frequency converter, op-amp version.

waveform is generally unimportant and usually is a pulse train. A voltage-controlled oscillator (VCO), as discussed in Chap. 11, performs a similar function except that its output waveform is sinusoidal or a symmetric square wave. Because a VFC can be used as an A/D converter, it is considered in this chapter.

Most VFC circuits involve an integrator which automatically resets itself when the output reaches a certain threshold. A suitable circuit is shown in Fig. 13-2. Application of a positive dc signal v_i to the input of the integrator (IC1) results in a negative-going ramp at the integrator output v' with a slope proportional to v_i. A comparator (IC2) connected to the integrator has an output v_o which is in negative saturation if v' is less negative than the threshold voltage V_T but rapidly switches to positive saturation at the point which the ramp v' becomes more negative than V_T. As v_o becomes sufficiently positive, the FET is switched on (resistance $< R_{on}$), causing C to discharge. Capacitor C' provides positive feedback and determines the output pulse length, which must be long enough to allow C to fully discharge ($C'R' > 5CR_{on}$). Depending on the type of comparator used, which in turn depends on the FET, an interface driver like that shown may be needed for digital compatibility. A narrow output pulse t_p is required for a reasonably high conversion linearity.

13-3   IC Voltage-to-Frequency Converters

An IC version of the VFC (Fig. 13-3) is based on a comparator-triggered one-shot. The standard version has two timing RC circuits, one charged from precision current source i_o and the other associated

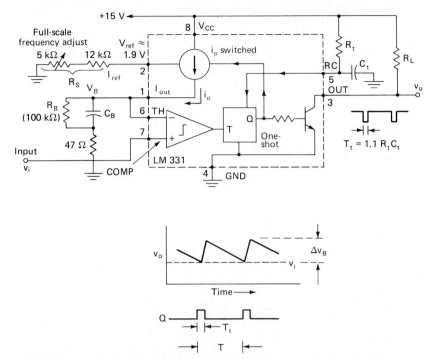

Figure 13-3 IC voltage-to-frequency converter, standard version.

with a one-shot multivibrator. The current source is switchable and is on when Q (one-shot output) is high.

Triggering of the one-shot (width $= T_t$) occurs when the comparator output is high, i.e., when $v_i > v_B$, and retriggering occurs if it remains high. In normal operation, after a starting transient, the one-shot delivers a standard charge $i_o T_t$, which charges the capacitor to the value $v_B = v_i + \Delta v_B$, where

$$\Delta v_B = \frac{i_o T_t}{C_B} \tag{13-1}$$

Next v_B decays (through R_B) relatively slowly until it reaches the comparator threshold value v_i, when the one-shot fires again. The period T required to decay by Δv_B (to v_i) is inversely proportional to v_B and thus (approximately) to v_i.

While the standard version of the circuit has the advantages of simplicity and operation from a single supply, the conversion accuracy is better than 0.1 percent. One disadvantage is a delay before the output frequency follows a step change in input voltage. Range is 1 Hz to 100 kHz with the components shown.

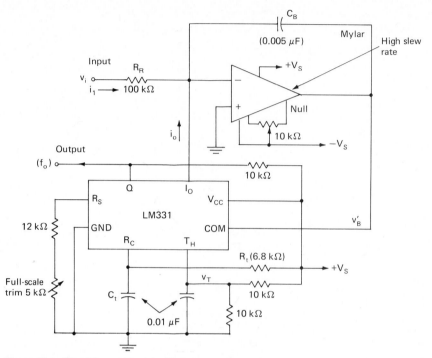

Figure 13-4 Precision voltage-to-frequency converter.

An improved VFC circuit (Fig. 13-4) uses an op-amp to linearize the charging of C_B. The input voltage v_i must be negative. Accuracy is improved because the summing point voltage is held to zero by the op-amp.

The inverse function, frequency-to-voltage conversion, is discussed in Chap. 14.

13-4 VFCs as A/D Converters

Utilization of VFC as an A/D converter requires the addition of a time interval generator and a pulse counter, as indicated in Fig. 13-5. The VFC produces a continuous train of pulses, and the number of pulses which pass through the NAND gate to the counter is determined by the timer pulse length T_{on}. A more detailed discussion of pulse counters is given in Chap. 16 and the design example sections. The number of pulses counted is

$$N = v_i K_f t_s \tag{13-2}$$

where K_f is the VFC conversion factor. In nearly all cases, the product $K_f t_s$ is chosen as an even power of 10 so that N is numerically equal to

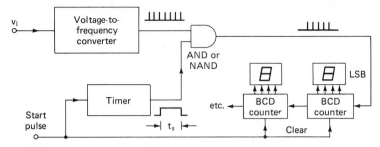

Figure 13-5 A/D converter (VFC version).

the input voltage with an appropriate placement of the decimal point. For example, if $t_s = 0.500$ s and $K_f = 200$, a 1.00-V input would result in $N = 100$ displayed on the digital readout. The display decimal point between the second and third digit would be turned on. Usually t_s is made slightly adjustable as a calibration control.

13-5 Ramp Converters

A simplified single-ramp converter is shown in Fig. 13-6. Before the start of the conversion cycle the integrator IC1 or ramp output v_B is at positive saturation. To initiate conversion, a positive pulse v_t is supplied at the "convert" or trigger input which causes the transmission gate to conduct and thus the capacitor to discharge so that v_B drops to zero. Simultaneously v_t clears the counters. When v_t returns to zero, the transmission gate is turned off and the ramp begins. While the input v_i is more positive than the ramp input v_B, the comparator output v_C is at **1**. When the NAND-gate input is high, pulses from the clock applied to the other input and of the NAND gate pass through to

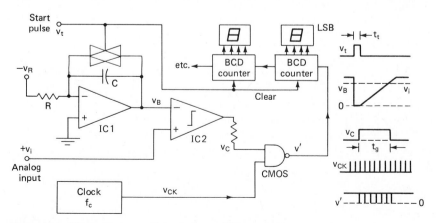

Figure 13-6 Block diagram of single-ramp A/D converter.

the counter. After a time interval t_g the ramp voltage v_B will exceed the input v_i and the counting will cease. It can be seen that the time interval will be proportional to the input voltage, provided, of course, that the ramp is linear, i.e.,

$$t_g = \frac{v_i\,RC}{v_R} \tag{13-3}$$

where v_R is a negative fixed reference voltage, which may be the supply voltage in less critical applications. During this time the number of pulses counted N is

$$N = t_g f_c \tag{13-4}$$

where f_c is the clock frequency, usually an even power of 10. With the proper choice of R, C, v_R, and f_c, it is possible to make N numerically equal to v_i with an appropriate placement of the decimal point.

As the discussion on the VFC emphasized, the pulse applied to the FET transmission gate (comparator pulse in this case) must be long enough to discharge the capacitor. Conversion accuracies of better than 0.2 percent are achievable with a stable clock frequency and a quality capacitor.

Dual-ramp (dual-slope) converters represent an improved version of the single-ramp converter. Conversion takes place in two phases. During the first phase (integrate or unknown) the integrator input is connected to the analog voltage v_i, as indicated in Fig. 13-7. During this phase the AND gate is held off, and a convert pulse resets the integrator (shorts the capacitor) and clears the counter. After predetermined time t_r, when the integrator output ramp $v_B = -V_m = -v_i t_r/RC$, the second phase (deintegrate or reference) begins by switch-

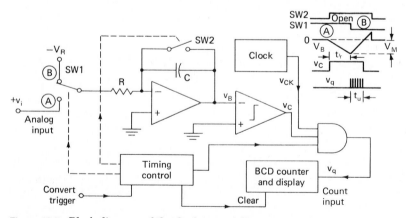

Figure 13-7 Block diagram of the dual-ramp A/D converter.

ing the integrator to V_R, which has an opposite polarity to the input v_i. Since the slope is now equal to V_R/RC, the time required for the integrator output to return to zero t_u will respond to $V_B = V_R t_u/RC$, and thus t_u is given by

$$t_u = \frac{v_i\, t_r}{V_R} \tag{13-5}$$

Note that the factor RC cancels, so that the capacitor need not be precision but should still be of high quality to avoid hysteresis. The comparator holds the gate on the time interval of t_u so that the number of pulses counted is

$$N = \frac{t_r f_c}{V_R}\, v_i \tag{13-6}$$

As before, N is proportional to v_i, and setting the constant of proportionality equal to an even multiple of 10 gives a readout directly in voltage.

If, as usual, t_r is derived from the clock frequency f_c by frequency division so that the count number $N_S = t_r f_c$ is known exactly, and N depends only on V_R, or

$$N = \frac{N_s}{V_R}\, v_i \tag{13-7}$$

The advantage of the dual-slope converter, then, is that the accuracy depends almost entirely on the reference voltage V_R, which can be precisely controlled with relatively little difficulty. Furthermore, the input voltage is averaged over the time interval t_r, so that the measurement is relatively insensitive to noise (with the single-ramp converter noise pulses near the end of the cycle can prematurely trigger the comparator).

Commercial digital voltmeters employ a large-scale-integrated (LSI) circuit dual-ramp A/D such as that shown in Fig. 13-8. Direct segment drive for up to a $4\frac{1}{2}$ digit (4 decimal digits and the leading digit 1) LCD or LED display is provided. Conversion takes place in three main phases. The first phase is auto-zero, during which the op-amp offsets are nulled. This is done by charging a (zero) capacitor C_Z with a voltage equal to the offset voltage.

The next two phases (integrate and deintegrate) are similar to the circuit of Fig. 13-7 except that the polarity of the reference can be reversed during the deintegrate phase if the input is negative. This occurs if the integrate output initially is positive, as sensed by the comparator at the end of the integrate phase.

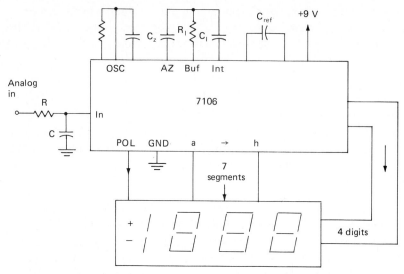

Figure 13-8 LSI digital voltmeter (A/D with multiplexed LED driver).

13-6 D/A Converters

Converting a digital signal to analog is quite simple, at least in principle, as shown in Fig. 13-9. The CMOS buffers act as a switch (see Fig. 3-9) which connects the output resistors (R, etc.) to the reference voltage V_R when the input is at **1** or to ground if at **0**.

Since the op-amp input V_- is at virtual ground, the current through the resistors $i = V_R/R$ is simply summed, and the output is

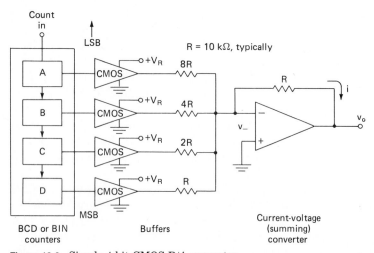

Figure 13-9 Simple 4-bit CMOS D/A converter.

$$v_o = -v_R \left(M_o + \frac{M_C}{2} + \frac{M_R}{4} + \frac{M_A}{8} \right) \tag{13-8}$$

where M_A, etc., are **1** or **0** depending on the logic state of each bit. Further bits are weighted as $R/10$, $R/20$, etc., for a BCD counter or as $R/16$, $R/32$, etc. for a binary counter.

Accuracy depends not only on the number of bits but on the precision of the reference voltage V_R and the resistors. Included in the total resistance must be the (on) resistance of the CMOS FET driver (typically 100 Ω). With an 8-bit BCD counter and 1 percent resistors ($R = 10$ kΩ) an overall accuracy of 2 percent is achievable.

To avoid a large number of different-valued resistors, a ladder network like that in Fig. 13-10a is often employed. The CMOS buffers are represented by a single-pole double-throw (SPDT) switch with V_R as the supply voltage. Circuit analysis is expedited by recognizing that each section can in turn be represented by the Thevenin equivalent of Fig. 13-10b. The result for this voltage-mode converter is

$$v_o = \frac{v_R}{2} \left(M_D + \frac{M_c}{2} + \frac{M_B}{4} + \frac{M_A}{8} \right) \tag{13-9}$$

Extension to 8 or more bits is straightforward. Commercial IC packages of 8, 10, and 12 bits are available, most with internal amplifiers, but an external reference V_R is often required. Some devices, termed *multiplying converters*, work with any polarity or magnitude of voltage

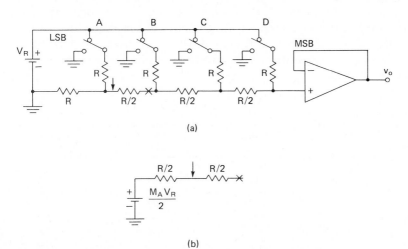

(a)

(b)

Figure 13-10 Ladder network (R–$2R$ type): (a) in a 4-bit D/A converter; (b) circuit equivalent.

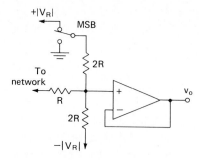

Figure 13-11 Polarity-conversion circuit for offset binary twos-complement code is implemented by the same circuit except that the MSB is complemented.

reference, including alternating current, and more properly can be thought of as a digitally switched attenuator (see design example E-5).

Polarity switching (bipolar output) can be implemented by either of the circuits of Fig. 13-11. The circuit must match the particular bipolar binary code, which is illustrated in Table 13-1 for 3 bits. With the binary-offset method, a fixed negative voltage equal to the weight of the MSB is summed with the ladder network output. When the MSB switches on, it just balances the negative offset and the output is zero, as can be seen from Table 13-1.

With the twos-complement method, an extra or sign bit (now MSB) is added to the code, which is equal to 1 for negative numbers. Many computers utilize this code because of the ease of manipulating negative numbers. The sign bit, if used, is a negative voltage equal to one MSB, which then subtracts from the sum of the other bits.

A third method of implementing polarity is to use the sign bit to reverse the sign of the output voltage. This can be implemented by a switchable inverter.

TABLE 13-1. Bipolar Binary Codes

Decimal number	Offset binary	Magnitude + sign	Twos complement	Relative voltage v_o
+3	111	011	011	+3/4
+2	110	010	010	+1/2
+1	101	001	001	+1/4
+0	100	000	000	0
−0	—	100	—	0
−1	011	101	111	−1/4
−2	010	110	110	−1/2
−3	001	111	101	−3/4
−4	000	—	100	−1

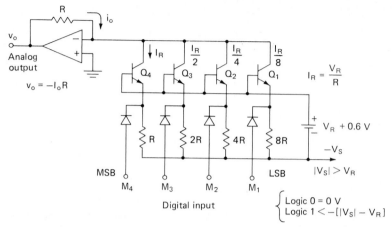

Figure 13-12 Binary-weighted BJT current sources as a D/A; $i_o = M_1I_R/8 + M_2I_R/4 + M_3I_R/2 + M_4I_R$.

Many higher-speed D/A converters are switchable BJT current sources. The currents are added by an analog summer (inverting amplifier), as indicated in Fig. 13-12.

13-7   Counter and Servo A/D Converters

A counter type of A/D converter is illustrated in Fig. 13-13. It uses a D/A converter, which allows comparison of the converter analog output v_A with the analog input v_i, assumed positive. Initially the counter is cleared by the convert pulse, so that $v_A = 0$ and the NAND gate allows clock pulses to pass. As time progresses, v_A increases in steps. At the point where v_A exceeds v_i, the comparator output switches negative and the pulses to the counter cease. At this point the digital output is essentially equal to the equivalent analog input.

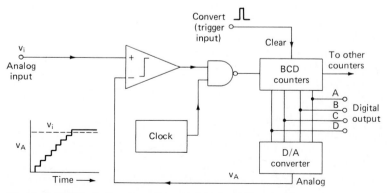

Figure 13-13 Block diagram of a counter-type A/D converter.

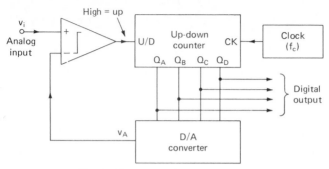

Figure 13-14 Block diagram of a servo-counter-type A/D converter.

Clock frequency determines the rate of conversion but has no effect on accuracy. While the illustration shows only 4 bits, this can be increased to any number although the time required grows rapidly with the number of bits, a significant drawback for many applications.

A related A/D converter (Fig. 13-14) is the servo type, so named because it follows or tracks the analog voltage like the servomotor with feedback. It is based on the up-down counter. Instead of turning off the clock pulses, the comparator reverses the counter direction so that it tracks the analog voltage. If the input voltage is constant, the counter will hunt or dither ±1 LSB about the proper value. Tracking is best for slowly changing input signal voltages and worst for large steps. Sometimes the clock and comparator output is transmitted to a remote counter, which follows the main counter. This method of operation is known as delta modulation.

13-8   Successive-Approximation
A/D Converters

Perhaps the most popular converter is the successive-approximation type (Fig. 13-15). It is similar to the counter type except that the bits are tested in succession, i.e., incremented in large steps starting with the MSB, rather than incrementing by the smallest step, as done when counting pulses. To do this, the bit is switched on during a test phase. If the bit causes v_A to exceed v_i as detected by the comparator, it is switched off and kept off; otherwise it is left on. Sequentially the control logic tests the next-most-significant bit until the conversion is completed. Clearly the process is much faster than counting pulses, especially if the number of bits is large. The main disadvantage is the rather involved control logic required.

Since successive-approximation converters are normally purchased as complete units (Fig. 13-16) or implemented by microprocessor

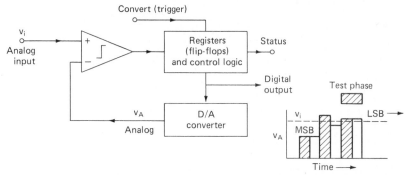

Figure 13-15 Block diagram of a A/D converter (successive-approximation version).

software, the circuit details will not be given here. A few converters require an external clock, which may be synchronized with other digital circuits, but usually the clock is internal as illustrated here. The conversion process begins immediately after a convert or start pulse is applied. During the time the conversion process (turning on bits and testing analog levels, etc.) takes place, a status or busy output goes to **1**. Digital output is not read until the status output returns to **0**. In some models the output does not change (is latched) until the conversion is complete. Serial output is sometimes provided as an option. The digital output often is microprocessor-bus-compatible, that is, can be disabled by a tristate driver (Sec. 20-7).

13-9 Flash Converters

A flash converter utilizes a series of fast comparators with the threshold inputs spaced at regularly increasing voltages. All comparators with threshold voltages below the analog input voltage v_i are

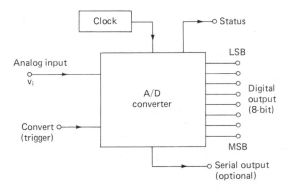

Figure 13-16 Block diagram of a A/D converter module.

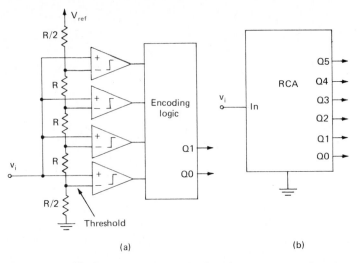

Figure 13-17 Flash converter: (*a*) simple 2-bit converter; (*b*) 6-bit converter.

turned on. Fast-encoding logic converts the highest on-comparator number to a binary code. Standard 6-bit flash converters (Fig. 13-17*b*) requires 64 comparators. The principle is related to the LED bar display driver (Fig. 5-23).

Unlike other A/D converters which utilize timing ramps or a series of sequential steps for the conversion process, the flash converter has only one step (voltage comparison) requiring appreciable time (the

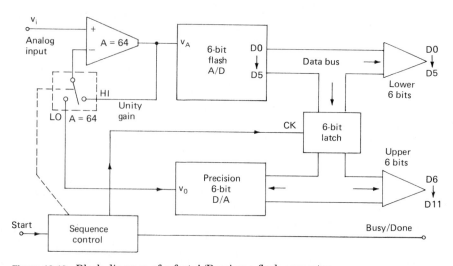

Figure 13-18 Block diagram of a fast A/D using a flash converter.

logic encoding is fast relative to the comparator response). Conversion speeds are in the 10- to 100-MHz range and are unapproached by other types of converters. Resolution is more limited than other types of A/D converters. Units can be cascaded (four 6-bit units comprise an 8-bit A/D).

High resolution at the expense of speed can be achieved by a two-step process involving a precision D/A converter and a difference amplifier (Fig. 13-18). Following a start conversion command, the sequence control switches the analog input to unity gain, and after a brief delay latches the resulting data, which is the upper 6-bits of the desired converted word. The output of the D/A is the analog equivalent of the input to a 6-bit accuracy. Unlike a standard D/A which has an accuracy of $\frac{1}{4}$ to 1 LSB, this precision D/A has an accuracy of better than $\frac{1}{64}$ LSB. Thus when the sequence control switches to the difference mode in the next step, the amplifier output is accurately equal to the difference between the analog input and the first 6-bits of the conversion process (amplified by 64). At this stage the flash converter output is equal to the low 6 bits of the converted signal, and the sequence control signals the end of the conversion process. Conversion speeds of 1 μs are typical.

14

Modulation and Demodulation

Often an analog signal of interest is proportional to the amplitude or frequency of a sine wave or other wave train. The process of extracting the signal from the ac wave is demodulation, and the reverse process is modulation. Included in demodulation is the conversion of an ac into a dc signal. Modulated signals may arise from sensors or sensor readout circuits. The purpose of demodulation then is to extract the desired signal. In other applications the signal is modulated to facilitate transmission to a remote site, where it is then restored by demodulation. Several signal-recording techniques also involve modulation and demodulation.

It is helpful to keep in mind that modulation and demodulation are inherently nonlinear and that the frequency content of the signal is transformed in the process.

14-1   Diode Rectifiers

Undoubtedly one of the simplest demodulation circuits is the diode rectifier. Transformer-coupled circuits, which have a dc return, require only the diode and filter network (Fig. 14-1). The principle is similar to that employed in power supplies (Chap. 21). It is necessary that the input voltage be greater than the diode drop (0.6 V for silicon diodes) for output current to flow. The input capacitor C_i removes ripple. If the carrier and modulating frequencies f_c and f_m are not widely separated, a sharp-cutoff filter, e.g., the Butterworth filter, may be required.

Capacitor coupling is preferred because the often bulky transformer is eliminated. A second diode (or a low-value resistor) is required to provide a dc ground return. Without D2 the coupling capacitor C_c

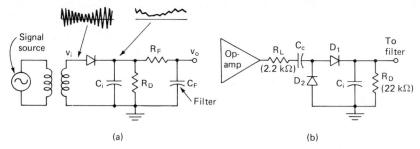

Figure 14-1 Diode AM demodulators: (*a*) transformer-coupled; (*b*) capacitor-coupled.

would charge to the peak signal voltage after an initial transient and no dc voltage would be developed across R_D since, of course, C_c does not pass direct current.

Diode demodulators have the advantage of simplicity but can be quite nonlinear, especially at low signal levels. They are suitable for noncritical, low-accuracy applications or at very high frequencies where other methods fail.

14-2 AC-to-DC Converters

A substantial improvement in performance over the diode rectifier is provided by the ac-to-dc converter. Where the output is a meter, specifically a microammeter calibrated in voltage, the circuit of Fig. 14-2*a* works well. It is a combination of the voltage-to-current converter (Chap. 2) and a bridge rectifier. Current i_2 from the op-amp output is identical to that flowing through the meter and R. Further, the magnitude of i_1 is equal to the current flowing through the meter; that is, $i_2 = |i_1|$ is the absolute value of i_1 or the rectified signal. The meter responds to the average value I_m of the current through it, which for a sinsusoidal signal is

$$I_m = \frac{2}{\pi} i_2 = 0.6366 |i_1| \tag{14-1}$$

For a sinusoidal input, $v_i = v_a \sin \omega t$, the meter current will be

$$I_m = \frac{0.6366 v_a}{R} \tag{14-2a}$$

or if v_i is calibrated in rms, where $v_{\text{rms}} = v_a/\sqrt{2}$,

$$I_m = \frac{0.9009}{R} v_{\text{rms}} \tag{14-2b}$$

It must be stressed that these factors are strictly true only for a pure sinusoidal signal.

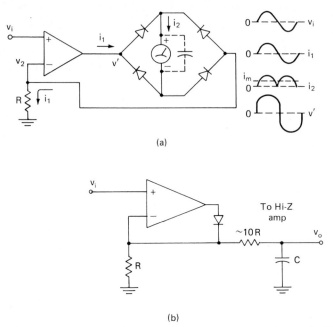

(a)

(b)

Figure 14-2 An ac-to-dc converter: (a) meter version and (b) half-wave version.

A major advantage of this circuit is that the diode drop ($\sim$0.6 V) is unimportant and good linearity is achieved even at low signal levels, down to the millivolt range. Diode or meter drop has a negligible effect because, as disscussed above, current feedback is employed and the meter current is forced by the op-amp to equal $v_i = v_2 = Ri_1$, provided the infinite-gain approximation holds. The op-amp output voltage v' is not important provided saturation does not occur. A disadvantage of this circuit is that the output is floating and cannot drive a subsequent stage (without a differential-input amplifier).

Often an ac-to-dc converter with a single-ended (dc) voltage output is required. For example, an ac-to-dc converter can be connected to a dc digital voltmeter so that the combination acts as an ac voltmeter. This circuit is also equivalent to an AM demodulator. A simple circuit is shown in Fig. 14-2b. A disadvantge of this circuit is that only a high-impedance load such as an op-amp input can be connected. Another drawback is that the rectification is half-wave with its high ripple.

A better circuit is the precision ac-to-dc converter circuit of Fig. 14-3. Its function can better be explained by considering the circuit with the capacitor C removed; the circuit then acts as a precision full-wave rectifier or absolute-value converter. For positive inputs, the current i_1

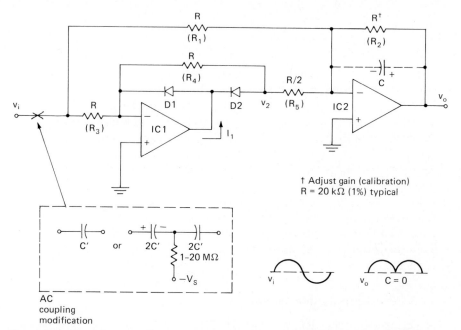

Figure 14-3 Precision ac-to-dc converter; without a capacitor C the circuit is absolute-value converter.

from IC1 passes through D2 (not D1), and thus IC1 acts as a unity-gain inverter with $v_2 = -v_i$.

The circuit acts as a full-wave rectifier or absolute-value converter. The term precision is applied because, unlike the circuit of Fig. 14-1, it is linear even for small signal inputs (smaller than the diode drop) as a consequence of the current feedback through the diodes.

Addition of a capacitor across R_2 causes IC2 to act as an active low-pass network or averager (Chap. 10). The output may be considered to be direct current but is more accurately thought of as a demodulated signal with frequency components lower than the break frequency ω_x, where $\omega_x = \frac{1}{2}\pi CR_2$.

Although for the purposes of discussion, it was assumed that $R_2 = R$, it is unessential to the circuit operation. Often R_2 is adjusted to provide an output calibration. In particular the dc output can be adjusted to correspond to either the peak or rms value of the ac input. By analogy with Eqs. (14-1) and (14-2) for a sinusoidal input, $v_i = v_a \sin \omega t$, the output v_o will be

$$v_o = \frac{0.6366 v_a}{R} \qquad (14\text{-}3)$$

Values of R_2 for peak and rms calibrations are

$$R_2 = \begin{cases} 1.55R & \text{peak} \\ 1.11R & \text{rms} \end{cases} \qquad (14\text{-}4)$$

Sometimes R_2 is made adjustable over a narrow range to allow calibration against a standard.

Without an input capacitor C' the circuit will respond to dc inputs, an advantage when measuring very low frequency signals but a disadvantage if the ac input signal has a dc component which is to be ignored. If a strictly ac reading is desired, an input capacitor C' is required. At high frequencies a (nonpolar) low-value electrolytic capacitor may be needed. Since the input direct current may be of either polarity, two back-to-back capacitors (each equal to $2C'$) are required to prevent reverse polarity. The junction is returned to a high negative potential through a high-value resistor to maintain the proper bias. This technique can be used whenever a large dc blocking capacitor is required.

14-3 Phase-Sensitive Detectors

A phase-sensitive detector (PSD), lock-in amplifier, or synchronous demodulator is similar in function to the ac-to-dc converter except that the output is determined by the phase of the input signal (with respect to a reference) as well as its amplitude. The principle is illustrated in Fig. 14-4, in which a SPDT analog switch is driven by a reference square wave. The RC network averages the switched signals and an instrumentation amplifier takes the difference. The input signal is here assumed to have the same frequency as the input signal but differs in the phase by an angle θ. A consideration of the waveforms at the various points in the circuit shows the average voltage on capacitor will be positive for $\theta = 0$, negative for $\theta = 180°$, and zero for $\theta = 0°$. The voltage across the other capacitor will be the negative of v_i and the output $(v_0 = \overline{v_1 - v_2})$ is $2v_i$.

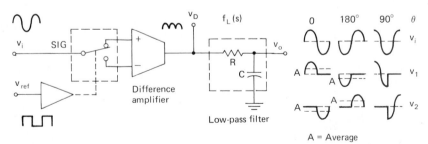

Figure 14-4 Block diagram of phase-sensitive detector or balanced modulator. Filters more complex than the single RC filter may be used.

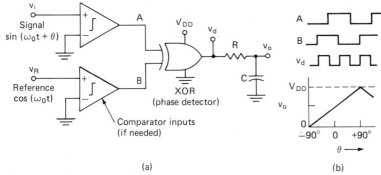

Figure 14-5 Digital (XOR) phase detector; (*a*) circuit with comparator drivers and (*b*) response.

The output of the PSD v_o for an input sine wave of phase θ relative to the reference will be

$$v_e = |v_i| K_p \cos \theta \qquad (14\text{-}5)$$

Phase detectors are basically the same device except that the signal amplitude is limited to a constant value and the phase is shifted by 90°, in which case the response is

$$v_d = K_d \sin \theta \approx K_d \theta \qquad (14\text{-}6)$$

A digital phase detector (Fig. 14-5) is constructed from an XOR gate. From the XOR truth table it can be seen that when $\theta = 0$ (A and B displaced 90°), the gate output V_d is at logic 1 for half the time and therefore the averaged output v_o is half the logic 1 level. Phase shifts θ about this point result in a proportional change in output, and therefore the XOR phase detector is inherently a linear detector (between $\theta = \pm \pi/2$). Note that $v_0 = v_{DD}/2$ for $\theta = 0$, resulting in a dc biased output response for $v_d = v_o - v_{DD}/2$.

14-4 Amplitude Modulation and Control

Modulation or control of the amplitude of an ac signal is ordinarily accomplished by a variable-gain or nonlinear amplifier. When the ac signal is a high-frequency, usually sinusoidal, carrier ($v_a \sin \omega_c t$) with an amplitude change proportional to the instantaneous value of a low-frequency modulating signal v_m, the composite signal v_s is considered to be amplitude-modulated:

$$v_s(t) = v_a A_m(t) \sin \omega_c t \qquad (14\text{-}7a)$$
$$v_s(t) = v_a[1 + m(t)] \sin \omega_c t \qquad (14\text{-}7b)$$

where $A_m(t) =$ instantaneous gain of modulator, $m(t) = K v_m(t) =$

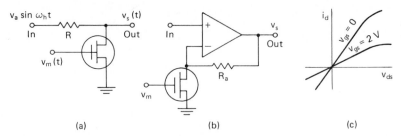

Figure 14-6 Amplitude gain control (FET versions): (*a*) attenuator; (*b*) variable gain; and (*c*) typical FET characteristics.

instantaneous modulation index [$0 < m(t) < 1$], and K = modulation sensitivity.

Discussions of amplitude modulation as found in texts on communications demonstrate that the modulation process generates two sideband frequencies, $\omega_c + \omega_m$ and $\omega_c - \omega_m$, which are present in addition to ω_c and ω_m.

For amplitude-control applications, A_m is adjusted so that the magnitude of the high-frequency signal is a specified value, and in this case A_m is not thought of as a modulation. Linearity here is usually not of prime importance. The FET gain controls of modulators shown in Fig. 14-6 depend on the variation of drain-source resistance R_{ds} with gate voltage V_{gs}. It should be noted that the FET acts as a pure resistance at low signal levels for signals of either polarity.

Although the FET provides a simple, high-speed method of varying gain, it can be quite nonlinear in two respects. The first type of nonlinearity is a curvature in the current-voltage characteristics (I vs. V_{DS} at higher voltages, typically over 0.5 to 2 V) which tends to clip the signal at higher (drain-source) voltage levels. This type of distortion can be made negligible by keeping the signal across the FET small.

The second type of a nonlinearity is the change of resistance with gate voltage. A sinusoidal voltage $v_m(t) = v_m(t) = v_m \sin \theta_m t$ applied to the gate will result in a distorted modulation. Such distortion can be avoided only if the modulation percentage is kept small or if corrective feedback is applied.

A gated amplifier is a particularly effective high-frequency modulator (Fig. 14-7*a*). A lower-frequency version (Fig. 14-7*b*) utilizes a variable transconductance (gain) op-amp.

14-5   Frequency Modulation

Frequency modulation (FM) and voltage-to-frequency conversion (VFC) are basically the same process, but the distinction in practice is

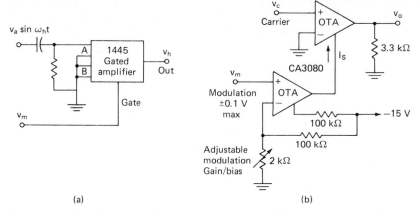

(a) (b)

Figure 14-7 Amplitude modulators: (a) gated-amplifier version; (b) variable-transconductance version.

that FM involves a comparatively small frequency shift Δf about some center frequency f_0, that is,

$$f = f_0 + \Delta f(t) = f_0 + K'v_m(t) \tag{14-8}$$

where K' is the modulation sensitivity. Frequency deviations $\Delta f/f_0$ of 5 to 0.1 percent are common.

Many of the VFC circuits discussed in Chap. 13 can be used as FM modulators. A suitable input would be a dc bias upon which is superimposed a relatively small ac modulating signal so that the output would be in the form of Eq. (14-8).

A high-frequency FM modulator based on a voltage-variable capacitor or varactor diode is comparatively simple to implement (Fig. 14-8). Voltage-variable capacitors are reverse-biased diodes made to maximize the high-frequency capacitance change associated with the variation of the depletion region with the voltage. Diodes act as capacitors

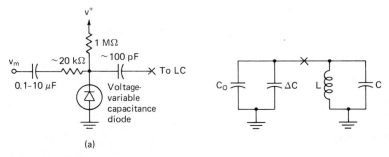

(a)

Figure 14-8 Voltage-variable capacitor modulator: (a) coupling circuit; (b) electrical equivalent.

best at high frequencies, above 1 to 10 MHz. The capacitance change per unit voltage is highest at lower dc bias voltages, but the nonlinearity also increases in this region. Care must be taken to avoid higher modulation amplitudes, which might drive the diode into the forward conducting region. Typically diodes have nominal capacitances (e.g., at -4 V) in the 10- to 100-pF range and vary 50 percent about this value with factor-of-2 variation in voltage.

The low-frequency modulation voltage must be coupled to the diode without shorting or excessively loading the high-frequency resonance circuit. Conversely the low reactance of the inductor at low frequencies cannot be allowed to short out the modulation-voltage source. The coupling network shown satisfies these requirements while also allowing an adjustment of dc reverse bias across the diode.

14-6   FM Detectors

A frequency-to-voltage converter based on the time averaging of narrow pulses acts as a low-frequency FM detector. A comparator and RC network of Fig. 14-9 triggers a monostable multivibrator at the input frequency. Its output is a continuous train of pulses of fixed width and height and the time average is proportional to frequency. Averaging is accomplished by an RC network ($R'C'$). For lower ripple, a high-order Butterworth filter is preferred. The output of the detector v_o is

$$v_o = K_v f \qquad (14\text{-}9)$$

The break frequency f_x of the filter is normally chosen to be at least an order of magnitude lower than the lowest frequency of the FM input.

A voltage-to-frequency converter IC such as the LM331 (see Fig. 13-3) can be reconfigured to act as a frequency-to-voltage converter. Actually only the one-shot section is used and the circuit is functionally identical to that of Fig. 14-9.

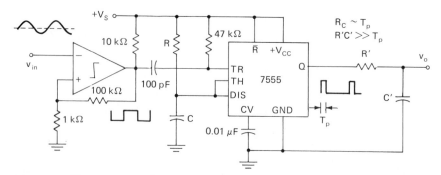

Figure 14-9 Frequency-to-voltage converter.

The phase-locked loop, an excellent FM detector, is discussed in the next section.

14-7 Phased-Locked Loops

Many modulation, detection, and frequency-synthesis methods, long recognized but difficult to implement, have become practical with the introduction of the IC versions of the phase-locked loop (PLL). Figure 14-10 shows a block diagram of this device.

The operation of the individual sections has been discussed previously. It consists of a phase detector (PD) connected to a voltage-controlled reference oscillator (VCO) which is automatically adjusted to equal the input frequency. By definition, lock occurs when the VCO frequency f_0 is equal to the input frequency f_i. The phase difference between the input and VCO is constant or a slowly varying function of time. Adjustment of the VCO to the input frequency f_i is accomplished by feedback from the PD output to the VCO input. The response time of the feedback circuit is determined by the low-pass-filter break frequency and the system gain. The demodulated output is proportional to the frequency deviation around the VCO center frequency f_{0o} since the VCO is designed so that its frequency shift is proportional to the control voltage. In other words, the PLL is an FM detector which follows the input frequency once lock is achieved.

The response of each section is summarized below. Adopting a common notation for PLL analysis, the PD output v_d in the linear region is

$$V_d(t) = K_d[\theta_i(t) - \theta_o(t)] = K_d\theta_e(t) \tag{14-10a}$$

where θ_1 = input signal phase, θ_0 = VCO phase, θ_e = difference or error phase, and K_d = detector gain or sensitivity.

Note that in general the phases are time-dependent, but when they are locked, θ must be small enough for the linear approximation to

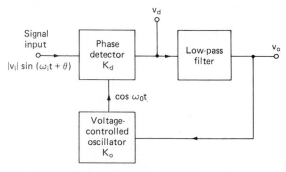

Figure 14-10 Basic phase-locked loop.

hold. Outside the linear region the response of a phase detector is usually approximated by

$$v_d(t) = K_d \sin \theta_e(t) \tag{14-10b}$$

The VCO frequency ω_0 is proportional to the PLL output voltage v_o around a center frequency ω_{o0}, or

$$\omega(t) = K_o v_o(t) + \omega_{o0} \tag{14-11}$$

When $v_0 = 0$, or the loop unlocks, the VCO frequency reverts to f_{o0} (if f_0 is substituted for ω_0, then $K_0' = 2\pi K_0$ replaces K_0).

The simplest low-pass filter is the single-pole response (Chap. 10) which, expressed as a phasor, is

$$\mathscr{F}_L(\omega) = \frac{v_o}{v_d} = \frac{1}{1 + j(\omega/\omega_x)} \tag{14-12}$$

where ω_x is the low-filter (or open-loop) break frequency.

By capture range is meant the range in frequency an initially unlocked loop will automatically lock onto or acquire the input signal. Shifts of the VCO frequency to that of the input can take anywhere from less than 1 cycle to millions of cycles (seconds or minutes to pull in or acquire lock). It is not obvious that an initially unlocked loop will ever spontaneously lock. When unlocked, the output of the phase detector is a distorted beat or difference frequency between the input and VCO frequency. Analysis (nonlinear) shows that the distorted beat frequency has a small dc component which is always in the direction to pull the VCO frequency toward the input frequency. If the filter has a long time constant (small system bandwidth), the magnitude of the beat frequency (including the dc component) will be small and there-fore the drift toward lock will be slow.

In the preceding analysis it was assumed that the VCO frequency f_o and input frequency f_i are equal at lock. Actually the loop will respond to any odd harmonic of the input frequency (f_i, $3f_i$, $5f_i$, . . .) because the phase detector (or PSD) responds to odd harmonics (but with decreas-ing sensitivity). Often the higher harmonics are outside the capture and/or hold-in range of the PLL, so that harmonic response is not a problem.

A practical IC PLL is pictured in Fig. 14-11. It is intended for operation at frequencies up to 30 MHz and therefore is suitable as an FM detector at standard radio broadcast intermediate frequencies of 4.5 or 10.7 MHz. Operation in the audio frequency range is possible although the required capacitors are somewhat large and the loop gain, which is proportional to operating frequency, becomes small. Like other RF amplifiers (Chap. 12), the input amplifier has a differ-

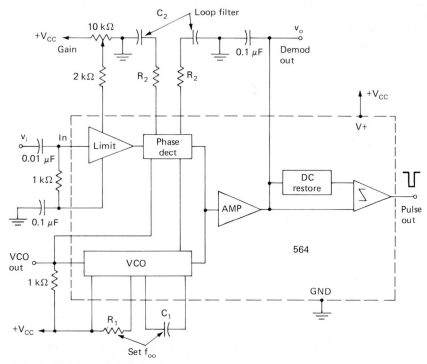

Figure 14-11 An RF-IF phase-locked loop (564) as an FM detector.

ential input which must be capacitor-coupled. For input amplitudes in excess of a threshold ($\sim$10 mV) the amplifier limits, and therefore the PD output becomes independent of, signal amplitude. The detector has a balanced output, like the PSDs described above, and therefore a dual low-pass filter ($\tau_1 = R_1 C_1$) is required. A capacitor C controls the center frequency f_{o0} of the VCO.

A characteristic of the VCO is that the fractional change in frequency $\Delta f / f_o$ with control voltage is a constant, independent of C and thus f_{o0}. This implies that the VCO gain K_o and thus the loop gain K_L is proportional to frequency, specifically $K_o = 0.05 f_o$ for the 564 PLL, assuming that the input voltage is high enough to provide limiting (in which case K_d is about 1.5 V/rad).

Other popular lower-frequency PLLs are the 565, XR 2212, and the 4046. Tone decoders and FSK demodulators (Chap. 19) also utilize the PLL.

Chapter

15

Noise and
Noise Reduction

At low signal levels the background noise present in every amplifier may be comparable to or greater than the signal voltage. Noise-reduction techniques can be divided into two groups: the design of low-noise amplifiers and signal enhancement by signal averaging or bandwidth reduction. Emphasis here will be placed on the latter techniques. Discussion of amplifier noise is presented here primarily to help in the informed selection and use of low-noise amplifiers rather than their detailed design.

It should be pointed out that the noise analysis assumes random noise rather than unwanted signals of other kinds (hum pickup or oscillation). Many of the random-noise reduction techniques actually reduce nonrandom noise as well.

15-1   Noise Sources and Terminology

Noise generation is inherent in any electric conductor as a consequence of the discrete nature of the charge carriers (electrons). Consider a resistor consisting of a conducting solid that contains randomly moving charge carriers. At any instant of time there is a certain statistical probability that the charge carriers will be concentrated more at one end than at the other, and as a consequence of this polarization a small voltage will develop across the terminals even in the absence of an external current or voltage source. This is referred to as thermal or Johnson noise. It is a basic statistical fact that the fractional deviation from the mean of a property decreases with an increase in the number of particles, so that the relative fluctuation of the charge carriers at either end will become smaller as the carrier

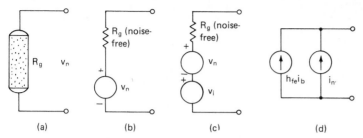

Figure 15-1 Noise equivalent circuits: (*a*) resistor; (*b*) equivalent of a resistor; (*c*) signal voltage source with internal resistance; and (*d*) transistor signal current source with shot-noise generator.

concentration, or conductivity, increases. In other words, the voltage fluctuation decreases with conductivity. A more detailed statistical analysis shows that the fluctuation expressed in terms of a thermal noise power P_{n0} is given by

$$P_{n0} = 4KT \, \Delta f \tag{15-1a}$$

where Δf = bandwidth within which measurement is made, Hz, K = Boltzmann's constant = 1.38×10^{-23} W · s/K, and T = absolute temperature, K.

The proportionality to the bandwidth occurs because statistical deviations are more likely in short time intervals and conversely tend to average to zero after a long time. When the noise power per unit bandwidth is the same at all frequencies, it is referred to as white noise. Expressed in terms of rms voltage v_{n0}, the terminal noise for an ideal resistor becomes

$$\bar{v}_{n0} = \sqrt{4KTR \, \Delta f} \tag{15-2a}$$

$$\bar{v}_{n0} = 1.28 \times 10^{-4} \sqrt{R \, \Delta f} \qquad \mu\text{V rms at 25°C} \tag{15-2b}$$

Since the bandwidth of any practical system is finite, the thermal-noise voltage is also finite and in fact small. For example a noise-free unity-gain amplifier with the extraordinary frequency response of dc to 1000 MHz ($\Delta f = 10^9$) when attached to an ideal resistor of 1000 Ω would have a noise of 128 μV. An equivalent circuit for a resistor consisting of a hypothetical noise-free resistor (as distinct from the ideal resistor) in series with a noise-voltage generator is shown in Fig. 15-1c.

Nonideal resistors and indeed all electrical components exhibit a noise voltage in excess of that of the thermal noise of an ideal resistor. The ratio of the noise power of a real component P_n to that of the ideal P_{n0} is defined as the noise factor F_n of the component.

$$F_n = \frac{P_n}{P_{n0}} = \frac{\bar{v}_n^2}{\bar{v}_{n0}^2} \tag{15-3}$$

where P_{n0} and v_{n0} are the thermal-noise power and voltage given by Eqs. (15-1) and (15-2). When the noise factor is expressed in decibels, it is referred to as a noise figure (NF)

$$\text{NF} = 10 \log F_n = 20 \log \frac{\bar{v}_n}{\bar{v}_{n0}} \tag{15-4}$$

Excess noise at all frequencies increases with the direct current through the resistor. A low-noise resistor (wire-wound or metal-film) may have a noise figure below 0.1 dB even when a direct current is present, but some carbon composition resistors have noise figures above 10 dB. Capacitor noise is uncommon.

15-2 Semiconductor Noise

Shot noise is a current noise associated with the flow of direct current across any junction. It is caused by the statistical fluctuation in the number of charge carriers passing the junction during a given time. For a reverse-biased diode, the shot-noise current $\overline{i_{ns}}$ is

$$\overline{i_{ns}} = \sqrt{2eI_{\text{dc}} \, \Delta f} \tag{15-5a}$$

$$\overline{i_{ns}} = (5.6 \times 10^{-10}) \sqrt{I_{\text{dc}} \, \Delta f} \tag{15-5b}$$

where I_{dc} is the direct current flowing through the junction and e is the electronic charge. In the particular case of a BJT the diode generating the shot noise is the base-emitter junction and I_{dc} is the quiescent collector current. The noise produced can be represented by a current generator or magnitude i_{ns} in parallel with the usual current generator $h_{fe}i_b$, as indicated in Fig. 15-1d. It can be seen that it is advantageous from the noise-reduction point of view to operate the transistor at as low a collector current as possible, taking into account the reduction of transistor-current gain h_{fe} and thus signal current as i_c is decreased. For many BJT input IC op-amps i_c is in fact quite low (in order to keep the input bias current low), and therefore the shot noise is not excessive. Shot noise is also present in vacuum tubes but is usually negligible in FETs.

At low frequencies the noise of transistors, or indeed all known amplifying devices, increases as the frequency decreases. This is referred to as $1/f$ (read "1 over f") noise since the noise power usually varies inversely with frequency. The low-frequency or $1/f$ noise P_{nl} is given by

$$P_{nl} = K \frac{\Delta f}{f} \tag{15-6}$$

where K is a constant for the device, Δf is the bandwidth as before, and it is assumed that $\Delta f << f$. In terms of voltage

$$\bar{v}_{nl} = \sqrt{\frac{K \Delta f}{f}} \tag{15-7}$$

Causes of low-frequency noise vary with the device, and there appears to be no fundamental reason why a device with very low $1/f$ noise cannot be constructed. Techniques of reducing $1/f$ noise have been successful, but no device completely free of this noise has been made. More generally this low-frequency noise varies as $1/f^n$, where $0.5 < n < 2$. In the case of the transistor (BJT and FET) the noise is associated with the random generation and recombination of surface traps at interfaces (gate or pn junction) which slightly modulate the current passing through the device.

Voltage reference diodes can be extremely noisy, especially at higher voltages. While they are often referred to as Zener diodes, actually only low-voltage diodes (below 4 to 10 V) undergo breakdown by the Zener mechanism: The distinction is important to noise considerations because the Zener mechanism exhibits only shot noise, while the noise due to the avalanche mechanism is orders of magnitude higher, including a substantial amount of $1/f$ noise. Higher-voltage reference diodes should not be used as bias or regulate the voltage of low-noise amplifiers. Noise in reverse-breakdown diodes may occur if the diode is overloaded or abused and is a sign of early failure.

15-3 Amplifier Noise

In this section the noise characteristic of amplifiers is discussed to aid in the selection of low-noise amplifiers, especially ICs, with due consideration to such factors as source resistance and frequency response.

Amplifier noise is always referred to the input. For example, if the noise at the amplifier output is 50 μV and the gain is 10, the effective noise with respect to the input is 5 μV regardless of the stage within the amplifier at which the noise actually originates. Of course, in designing an amplifier it is necessary to know at what point the noise originates, and in any well-designed amplifier the noise of all stages other than the input stage is made negligible. An equivalent circuit for a high-input-impedance amplifier connected to a (low-level) signal source v_i with an internal (Thevenin) resistance R_g is shown in Fig.

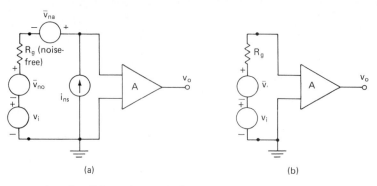

Figure 15-2   Amplifier noise equivalent circuits (grounded input).

15-2a. For a differential-input op-amp it is more accurate to apply the equivalent circuit shown to each input separately, but in practice the noise of one input is usually negligible because it is connected to a low-impedance source and/or can be referred (added) to the other input so that the simpler single-ended input model or equivalent is usually adequate. Here the $\overline{v}_{no}$ is the magnitude of the thermal noise associated with the source resistance R_o, $\overline{v}_{na}$ is the amplifier noise voltage (at a particular frequency) referred to the input, and i_{ns} is the noise-current generator due in part to shot noise coupled (directly or capacitively) to the input. The noise current is small but not zero for FET amplifiers.

Note that even in the simpler equivalent circuit, two noise-voltage sources and one noise-current source are present. It is convenient to combine them into a single noise-voltage source (Fig. 15-2b) with a total voltage of

$$\overline{v}_n = \overline{v}_{na} + \overline{v}_n + \overline{i}_{ns} R_g \tag{15-8}$$

The plot of noise voltage as a function of source resistance (Fig. 15-3a) illustrates the increase in noise with resistance. Alternatively the noise figure of the amplifier, obtained by inserting the noise voltage expressed by Eq. (15-8) into (15-4), can be plotted as a function of source resistance. It expresses the contribution of amplifier noise related to that of the unavoidable (once R_g is fixed) thermal noise of the source.

A more definite measure of amplifier noise is the signal-to-noise ratio S_n in decibels.

$$S_n = 20 \log \frac{v_i}{v_n} \tag{15-9}$$

It should be noted that a 3-dB noise figure achievable in good but not

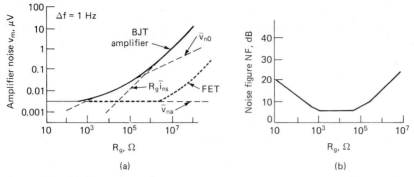

Figure 15-3 (a) Typical amplifier noise; (b) noise figure as a function of source resistance.

extraordinary low-noise amplifier represents a noise voltage only 41 percent above the irreducible thermal noise. Only infrequently would the effort or cost of a very low noise amplifier (0.1 to 1 dB) be justified.

It is important to construct the amplifier with proper shielding, grounding, and low-ripple power supplies if the injection of unwanted signals, often referred to as noise, is to be avoided. Placement of the preamplifier in a separately shielded box near the signal source and remote from hum-producing transformers is suggested (Fig. 15-4). Rigid mounting of components helps provide immunity to room vibrations. A separate ground return to the signal source avoids noise pickup on the ground line. Ripple and noise from standard regulated

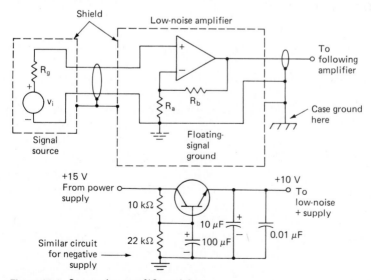

Figure 15-4 Low-noise amplifier wiring.

power supplies may be too high for low-noise applications. Batteries with bypass capacitors are good low-noise power sources. Alternatively the voltage may be filtered by a low-pass RC filter to remove noise and ripple.

15-4   Bandwidth Limitation

Once a low-noise amplifier and the optimum source resistance have been selected, in general, the only further means of reducing noise is to reduce the amplifier bandwidth since, as pointed out above, all random-noise voltage is proportional to the square root of bandwidth [Eqs. (15-2), (15-5), and (15-7)]. It is advantageous in particular to limit the low-frequency response in order to minimize $1/f$ noise. Commercial low-noise preamplifiers often have adjustable high- and low-pass filters to allow limitation of low- and high-frequency response of the amplifier to that just required to pass the signal without distortion or attenuation. Excess bandwidth increases noise without affecting the signal.

The filters which reduce the bandwidth need not be in the input amplifier. Often it is most convenient to limit the bandwidth in the second stage, where the signal level is much higher than any noise generated by the components making up the filter but not so high that the amplitude noise outside the desired bandwidth can drive the amplifier into a nonlinear region (on peaks, in particular) before being taken out by the filter. Suitable filters are discussed in Chap. 10.

Under conditions where the signal frequency is nearly constant the bandwidth can be made very narrow and the noise greatly reduced correspondingly. Narrow-band amplifiers can be used to limit the bandwidth, but if the ac signal is to be demodulated, as is usually the case, the PSD (see Chap. 14), which combines the bandwidth-limitation and detection functions in one circuit, is superior in terms of both performance and ease of design. Where the noise level is high, however, it may be advantageous to precede the PSD by a narrow-band amplifier to limit the noise peaks and thus reduce the linearity and dynamic-range requirements of the PSD.

As an example of PSD as a narrow-band detection system, consider the circuit of Fig. 15-5, which is intended to measure very small changes in resistance from the nominal value R_0. Many transducers or precision bridge measurements can use this or a similar circuit. As indicated in the discussion of the bridge amplifier, the output voltage is proportional to the bridge unbalance ΔR. Thus, the dc output voltage is also proportional to ΔR. The output frequency response of the PSD ($f_x = \frac{1}{2}\pi RC$) is determined by the time constant of the output low-pass network, which can be made arbitrarily low by choosing sufficiently

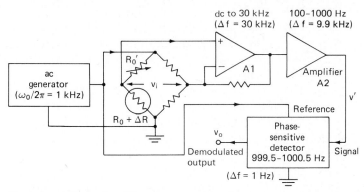

Figure 15-5 Example of the phase-sensitive detector as a narrow-band amplifier and demodulator.

large capacitors. In other words, only input frequencies between $f_0 + f_x$ and $f_0 - f_x$ after demodulation are passed by the filter, resulting in a bandwidth $\Delta f = 1$ Hz. Note that the noise can be made very small by increasing the PSD time constant.

Any signal which does not bear a fixed phase relation to the reference wave, including hum or other undesirable signals, has a zero average and therefore will be rejected by the PSD. A narrow-band amplifier will also reject unwanted signals but has the disadvantage of being difficult to tune, especially if the bandwidth is very narrow. By contrast a diode rectifier or converter responds to noise and unwanted signals as well as the desired signal, and subsequent filtering is not effective in bringing out the signal.

15-5 Signal Averaging

An effective means of noise reduction for periodic or repetitive signals is signal averaging. Each repetition of the signal is equivalent to that seen as one sweep of an oscilloscope, and the averaging process can be thought of as taking the time average of many sweeps at each point along the time axis. The assumption here is that random noise or any unwanted signal not synchronized to, or correlated with, the desired signal will tend to average to zero over a long time while the desired signal will remain unchanged. An average $\tilde{S}_n$ of n repetitions of sweeps of the noisy signal-to-noise ratio S_n is given by

$$\tilde{S}_n = \frac{S_n}{\sqrt{n}} \tag{15-10}$$

One way of interpreting this result is to think of the averaging process as increasing the effective period over which each point is taken, a procedure equivalent to reducing the bandwidth. This signal averaging in effect reduces noise by decreasing the system bandwidth. An average over $n = 10^4$ sweeps resulting a signal-to-noise improvement of 100 times (40 dB) is not uncommon.

Signal averaging is often done by a computer after A/D conversion. However, the averaging often requires that a large number of bits per second over an extended time be processed, and the dedication of a substantial portion of the capacity of a general-purpose computer to this simple task may be economically unsound. Even when a computer is available, it may be wise to preprocess the data with a specialized signal averager or microprocessor before transfer to a larger computer for further processing.

SIGNAL AMPLIFICATION AND PROCESSING DESIGN EXAMPLES

Example 4   Fourth-Order Butterworth Low-Pass Filter (4-kHz cutoff)

DESCRIPTION: The filter attenuates signal frequencies above breakpoint sharply (80 dB per decade) while not appreciably attenuating lower frequencies.

SPECIFICATIONS
    Voltage gain: 2.58
    Breakpoint (-3 dB):   4.0 kHz*

DESIGN CONSIDERATIONS: A fourth-order filter is made by cascading two quadratic filters following the design procedure outlined in Sec. 12-7. For simplicity in obtaining capacitors, the equal-value-capacitor version of the op-amp configuration (refer to Fig. 12-16a) was selected. A standard-value capacitor which has a reactance of the order of 10 kΩ (optimum impedance range for op-amps) was chosen. From the resonance condition the value of the resistors ($R = \frac{1}{2}\pi FC = 8.06$ kΩ) was then calculated. Butterworth filter constants b are 1.845 and 0.765, according to Table 11-2 so that the feedback resistors R_1 and R_2 are 0.16R and 1.24R, respectively. Gains of the two stages are 1.16 and 2.34, respectively, for a total gain of 2.58. A unity-gain input buffer is required if the filter is not driven by a low-impedance source, e.g., another op-amp stage.

---

* Attenuation (relative to dc) at 3 and 6 kHz is 0.4 and 26 dB, respectively.

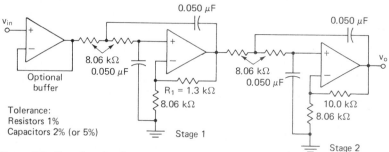

Figure E-4   Fourth-order low-pass filter.

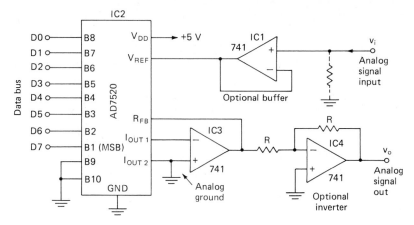

Figure E-5 Digital gain-controlled amplifier.

Example 5 Amplifier with Digital Gain Control

DESCRIPTION: The analog amplifier has a gain which is determined by an 8-bit digital signal.

SPECIFICATIONS
Voltage gain: 0 to 1.00 in steps of 1/255
Frequency response: dc to 1 MHz
Analog voltage range (input-output): ±10 V
Digital control: 8-bit TTL level (0 to 4 V)

DESIGN CONSIDERATIONS: A standard multiplying D/A converter based on a digitally switched ladder network (Fig. 15-10) acts as a variable resistor in an inverting-amplifier configuration. The signal, which may have either polarity, is applied to the reference-voltage input. Internal (IC2) feedback resistors provide unity gain for the inverting amplifier (IC3). An output amplifier (IC4) reinverts the signal if an uninverted amplifier is required. The input amplifier (unity gain) is required only if the signal source does not have a low impedance.

Standard TTL-level signals constitute the 8-bit digital data bus, which can be connected directly to a microprocessor output port. All bits equal to 0 correspond to zero gain, and all bits equal to 1, maximum gain.

SIGNAL AMPLIFICATION AND PROCESSING DESIGN PROBLEMS

16 Draw a circuit which will produce an output (inverted acceptable) equal to the slope of an input ramp signal v_i. The calibration should be such that a rise of 1 V/ms results in a 1-V output. Describe the output which occurs on the practically instantaneous return slope.

17 Design a notch filter to reject 60 Hz while passing all other frequencies from dc to 100 kHz. Specifications include a dc gain of 10.0, an input impedance over 100 kΩ, and an output impedance under 10 Ω. If desired, calculate bandwidth.

18 A bandpass filter is desired which will pass 2 kHz (−3 dB points at about 1.6 and 2.4 kHz) with unity gain.

19 Design an EEG amplifier filter to pass frequencies between 4 and 16 Hz (-3 dB points) while sharply attenuating signals outside this band. In particular, the gain at 60 Hz should be at least 40 dB below the gain at 8 Hz (midband). A midband gain of 1000 is needed, and the dc offset voltage at the output should be below 0.1 V. *Hint*: Use high- and low-pass Butterworths.

20 Draw a circuit diagram of sinusoidal oscillator with an output of 20 V peak-to-peak at a frequency of 400 Hz.

21 Design a wideband amplifier with a gain of 10.0 from dc to 10 MHz.

22 Design an A/D converter (8-bit) with a range of 0 to 1 V dc based on either the single-ramp, counter, or servo method. A conversion time of 0.1 s is acceptable.

23 Design an ac-to-dc converter-type demodulator for a strain-gage bridge excited by ac (1 kHz). Assume the output of a bridge amplifier at maximum strain ($E = 0.01$) is 5 V peak-to-peak. A dc component may be present due to an uncompensated amplifier offset. An output of 1.0 V at full scale is required. Frequency components of the strain (modulation of the 1-kHz carrier) will not exceed 20 Hz. A PSD is suggested.

24 Design a PLL which will demodulate pulse-code modulation. The center frequency is 2.0 MHz and input amplitude 10 mV. The frequency deviation for a pulse (**1**) is 10 kHz, the pulse clock frequency is 200 Hz, and at least one **1** (and one **0**) pulse is transmitted every 10 clock cycles. The circuit output pulses must be CMOS-compatible (0 to 5 V). *Hint*: Use an ac amplifier to bring up the output level to the point where a comparator or Schmitt trigger will work well, possibly using a diode clamp to restore dc.

Data Switching, Control, and Readout

16

Pulse Timers and Counters

The various standard circuits for pulse timing, delay, counting, and decoding described here are major or minor subsections of numerous electronic instruments. Most are based on simpler digital devices described previously. The related pulse-generation circuits were discussed in Chap. 11.

16-1 IC Timers

IC timers like the popular 555 timer of Fig. 16-1 are versatile devices which can produce a single pulse of known duration or a continuous pulse stream. Operation can be understood by examining the timer connected as a triggered one-shot. Internally the timer is basically an RS-type flip-flop with comparators on the set (S) and reset (R) inputs. An extra open-collector output is provided for external capacitor discharge. The comparators have internal thresholds fixed at two-thirds and one-third the supply voltage V_{CC}, except for a few applications where an external voltage is injected into the control-voltage input. Note that the pin labeled "threshold" or THRES is somewhat misleading since usually the fixed-voltage input to a comparator is thus named. The discharge of an external capacitor by the transistor occurs when the flip-flop output Q is at 0.

In the normal reset or rest state, before the arrival of the trigger pulse, the output Q is low and the discharge transistor holds the capacitor voltage at zero. Since the capacitor is connected to the THRES input, $\overline{R}$ is high and R is low. It is necessary that the TRIG input be more positive than $\frac{2}{3}V_{CC}$ in this state because if it were not, S and therefore Q would be high.

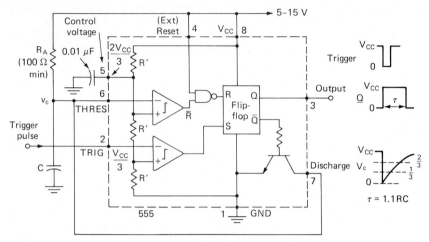

Figure 16-1   The 555 timer as a one-shot (monostable) multivibrator.

Triggering of the timer requires that the input pulse drop from its positive rest value (above $\frac{2}{3}V_{CC}$) to a value less positive than $V_{CC}/3$. The required pulse is sometimes referred to as a negative-going or (inaccurately) a negative-edge-triggered pulse. When this occurs, S and therefore Q go high. Also at this time the discharge transistor turns off, and the capacitor C starts charging. As the capacitor voltage rises to the upper threshold value $\frac{2}{3}V_{CC}$, the flip-flop is reset (Q low), ending the timed period (provided the trigger pulse is no longer present). If the trigger pulse is less positive than the lower threshold $V_{CC}/3$, the flip-flop remains set (Q high).

The time τ required for the capacitor to charge to $\frac{2}{3}V_{CC}$ corresponds to $1.1R_AC$. Note that if R_A is too small, the current flow through the discharge transistor will be excessive. An external reset ($Q = 0$ for RESET $= 0$) is provided for special applications. External variation of timed period is accomplished by applying a signal to the control-voltage input. Capacitors across the power-supply terminal (0.1 to 5 μF) and perhaps from the output to ground may be required to eliminate multiple pulses (glitches).

Timing accuracy depends on the resistor and capacitor stability. With Mylar capacitors the accuracy will be better than 1 percent, and reproducibility from pulse to pulse is about 0.1 percent. Higher-precision timers, for example, the LM322/LM3905, are at least a factor of 10 more reproducible than the 555. However, accuracy still depends on capacitor and resistor stability, which are likely to be the limiting factors. If high timing stability ($<$0.2 percent drift) is required, a crystal-controlled timer is a better choice.

Connection of the 555 timer as an astable multivibrator or continuous-pulse generator is discussed in Chap. 11.

16-2 Digital One-Shot Multivibrators

One-shot multivibrators are also available in both CMOS and TTL digital series. They feature high speed and direct compatibility with digital systems but have the disadvantages of a fixed threshold and hysteresis. Also long pulse widths (>10 s) require inconveniently large capacitors. Standard and retriggerable units are shown in Fig. 16-2. Both are triggered when the input trigger pulse reaches an upper threshold level and cannot be retriggered unless the trigger input drops below a lower threshold and again rises to the upper threshold. The retriggerable type differs only in that the output pulse width is measured from the time of the last trigger input, as indicated by the pulse-test sequence of Fig. 16-2. Output pulse width τ is determined by the product of R_A and C_A. Variable widths are achieved by making R_A adjustable.

Often a variable trigger level (or hysteresis) is desired, and in this case a comparator can be used preceding the one-shot. The threshold is adjusted by the voltage on the noninverting input.

For applications where timing accuracy is not critical, the simple CMOS one-shots of Fig. 16-3 may suffice. They have the advantage of requiring only inexpensive standard gates, possibly unused sections of an existing package, rather than a special IC. The Schmitt-trigger type is based on the discharge of an RC network charged by the trigger pulse. Note that the output Q rises on the leading edge of the trigger pulse but the timing starts on the falling edge. The circuit therefore

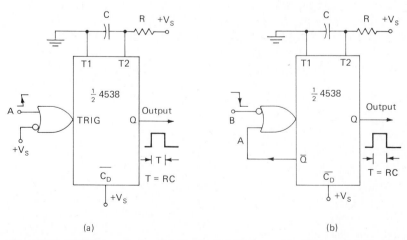

(a)   (b)

Figure 16-2 CMOS one-shot multivibrator: (*a*) retriggerable; (*b*) nonretriggerable.

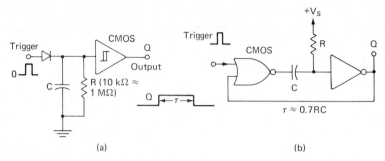

Figure 16-3 One-shots built from CMOS gates: (*a*) Schmitt trigger, pulse-stretcher type; (*b*) edge-triggered.

functions as a pulse stretcher. It is necessary for the input pulse to last long enough to charge the capacitor. Long output-pulse durations (at least 10 s) are practical. Reversal of the diode polarity and return of R to $+V_{CC}$ instead of ground allows triggering by an inverted trigger pulse (negative logic).

A one-shot which is triggered by a positive edge is shown in Fig. 16-4*b*. Positive feedback is employed to sharpen the transition. Timing is again determined by the RC product. TTL-type one-shots are also available (Fig. 16-4). They are useful for generating shot pulses (down to 50 ms) but work poorly for longer pulse widths.

Timers can be made from counters by disabling the counter input at a given count (Fig. 16-5).

16-3 Triggered Sweep

A triggered sweep generats a ramp or linearly arising voltage with time following an input trigger pulse. An integrator with a constant-voltage input is employed to produce the sweep. The circuit of Fig. 16-6 has a flip-flop, set by the trigger signal. The output of the flip-flop v_o is integrated (IC1) to form the ramp. When the ramp reaches the

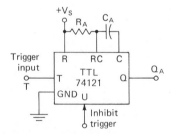

Figure 16-4 One-shot multivibrators, TTL version.

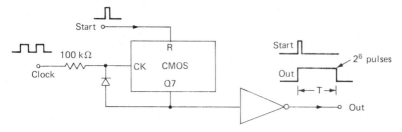

Figure 16-5 Counter-type timer.

threshold level, the comparator (IC2) output goes positive, resetting the flip-flop and the integrator (discharge of C through the FET). The ramp slope $S_R = dv_o/dt$ where V_S is the op-amp output voltage at saturation.

$$S_R = \frac{V_S}{RC} \tag{16-1}$$

Termination time of the ramp can be made variable by adjusting the ramp threshold resistor R_T. In this case the pulse length T_A taken at the alternate output V_o' varies linearly with the threshold potentiometer setting, and the circuit acts as a one-shot multivibrator.

16-4 Pulse-Delay Circuits

Pulses must often be delayed in order to perform a series of operations in the proper sequence. In most cases the width of the delayed pulse need not be the same as that of the input pulse, and in this case the circuit of Fig. 16-7 would be suitable. The delay time τ_D is determined by the length of the first timer connected as a one-shot multivibrator. Because the 555 timer is triggered on a negative-going edge, the second timer will be triggered only at the end of the pulse v' produced

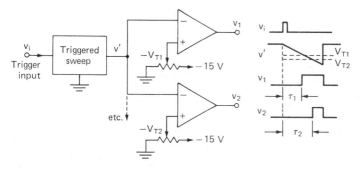

Figure 16-6 Triggered sweep.

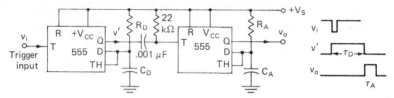

Figure 16-7 Dual one-shot pulse-delay circuits.

by the first timer. Output pulse width τ_A is controlled by the time constants R_A and C_A of the second timer.

The more elaborate pulse-delaying circuit shown in Fig. 16-8 is especially useful if several delayed pulses v are to be generated. It consists of a triggered sweep followed by one or more comparators. The output goes positive as the threshold V_{T1}, V_{T2}, . . .) is reached. Time delay is calculated from the relation

$$\tau_1 = \frac{V_{T1}}{S_R} \tag{16-2}$$

where V_{T1} is the magnitude of the threshold voltage and S_R is the ramp slope [see Eq. (16-1)]. Similar expressions hold for τ_2. Note that the threshold voltages are linearly dependent on the potentiometer settings.

Where the length of the delay is not critical, the CMOS circuit of Fig. 16-9a may be adequate. The output pulse is delayed by the RC network. Positive feedback is provided to reduce the CMOS power dissipation near the transition point.

Sometimes a dual-pulse circuit is desired in which one of the pulses is delayed slightly. One application is the latch-and-clear sequence needed when transferring digital data from a counter to a latch. The

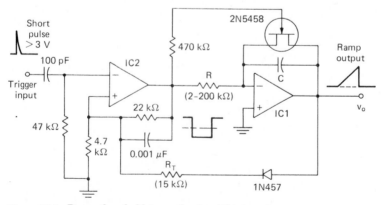

Figure 16-8 Ramp-threshold type of pulse delay.

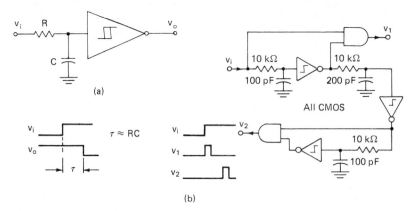

Fiugre 16-9 Simple CMOS pulse-delay networks: (*a*) single delayed output (inverts); (*b*) dual output pulses, one delayed.

circuit Fig. 16-9*b* is intended for this purpose. It provides an 1-μs pulse v_1 immediately following a step input and a second 1-μs pulse v_2 about 2 μs later.

16-5 Bounceless Switches

When a pulse sequence is initiated by a push-button switch or relay, it is usually necessary to add a special circuit to eliminate contact bounce or jitter, which can be interpreted by logic circuits as multiple pulse inputs. Contact bounce generally lasts for a few milliseconds at most, and therefore a circuit which has a time constant of 20 to 50 ms is effective in reducing the problem. The circuit (Fig. 16-10) produces a pulse Q which lasts as long as the switch is depressed.

The second type or circuit (Fig. 16-11), popular in microprocessor applications, consists of an *SR* flip-flop which is set by the first pulse

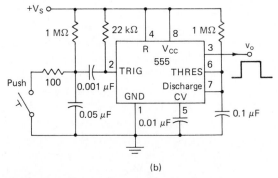

(b)

Figure 16-10 Bounceless switch, timed version.

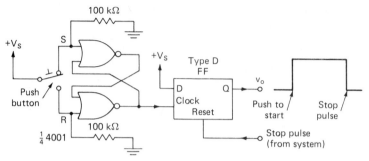

Figure 16-11 Bounceless switch, flip-flop version.

from the SPDT push-button switch and reset by a stop pulse returned by the system after the desired series of operations has been completed. Note that bounce or multiple pulses to one input of the SR flip-flop do not produce multiple output pulses of the SR since bounce is associated with only one switch contact at a time. If the switch is still depressed at the time the stop pulse occurs, the type-D flip-flop will reset but will not set until the switch is pushed again.

16-6 Pulse Counters

Frequently the number of pulses occurring within a period must be known. The circuit of Fig. 16-12 is intended to count the number of pulses between an externally supplied start pulse S and a stop pulse R. An AND gate connected to the flip-flop output controls the flow of pulses into the counters. A NAND gate can replace the AND gate since the counter in either case requires one full clock cycle per count. A clear pulse is generated on the leading edge of the gate-control pulse. For TTL the pulse-differentiator resistor R_p must be low enough to sink the currents flowing out of the clear inputs (see Chap. 3). Of course, if a gate-control pulse with a width equal to the timing interval T_c is available, the flip-flop is not needed.

The counter and display sections are indicated in terms of a block diagram because circuit details have been described elsewhere (Chaps. 3 and 5). A specific design example is given in Example 2.

Often counters employ latches (described in Chap. 17) to hold the final count for display (Fig. 16-13). In this case, the counter can be cleared immediately after the data have been stored in the latch. Latching is controlled by the latch-enable pulse, which usually occurs immediately following turnoff of the counter. The clear pulse must of course follow the latch pulse. The delay may be brief, as generated by a pulse-delay circuit (see Fig. 16-9), but often the clear pulse occurs instead on the rising edge of the counter-edge pulse.

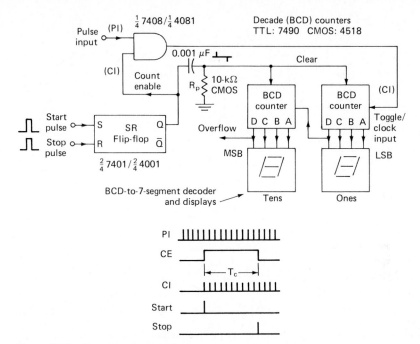

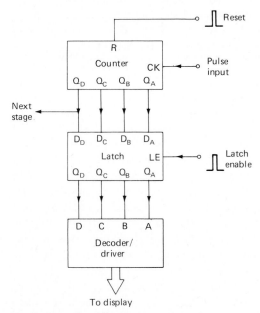

Figure 16-12 Decade pulse counter.

Figure 16-13 Decade counter with latch.

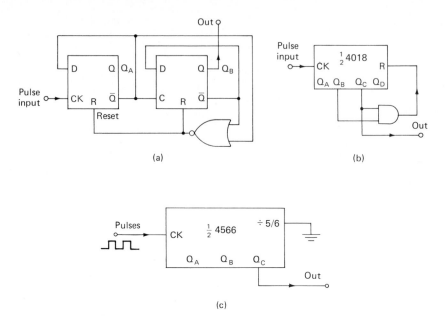

Figure 16-14 Counters: (*a*) base 3; (*b*) base 6; (*c*) industrial-type base 5 or 6.

Sometimes counters based on a scale other than 2 (binary) or 10 (decade) are desired, and in this case external logic can be used to provide the reset. As examples the scale of 3 and 6 are suggested by Fig. 16-14. Such counters find application in frequency dividers (see Section 16-7).

16-7 Frequency Dividers and Multipliers

A number of simpler frequency-divider circuits have been discussed previously. These include divide by 2 (Fig. 3-9), divide by 10 (Fig. 3-11), divide by 3 and divide by 6 (Fig. 16-14), and divide by $2N$ (Fig. 3-10). Division by multiples of these numbers, for example, 100, 20, 60, can be accomplished by cascading several stages. For applications where the division is by a large number which is not a multiple of a standard divider or is variable, it is best to use a general-purpose divide-by-N counter. Either counter specifically designed as a divide-by-N counter, for example, the 4522, or a presettable down counter (Fig. 16-15) is suitable.

The cycle of a divide-by-N counter starts when the individual counters are simultaneously preset by the application of a pulse to the preset-enable (PE) inputs. Each counter stage corresponds to one (decimal) digit (n_1, n_2, etc.) of the number N. For example, if $N = 93$, counters 1 and 2 are present to $n_1 = 3$ and $n_2 = 9$. Note that $n_1, n_2, \ldots$

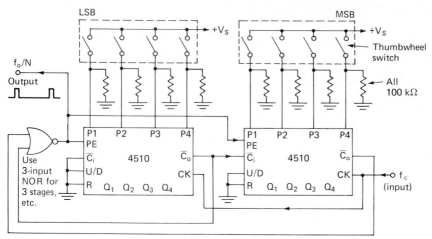

Figure 16-15 Two-stage divide-by-N decade counter, expandable.

are actually 4-bit binary numbers (here BCD), which are supplied by thumbwheel encoders or, in more complex circuits, by a set of registers or latches. The first time counter 1 counts down, it will require n_1 clock pulses before it reaches zero count, at which point the carry output $\overline{C}_o$ will go to **0**. Subsequent counter-1 cycles will require 10 counts to produce a carry $\overline{C}_o = \mathbf{0}$. The carry out $\overline{C}_o$ of one stage is connected to the carry input $\overline{C}_i$ of the next. Actually the $\overline{C}_i$ acts as a counter enable, and the stage will count only if the $\overline{C}_i$ of that stage is at **0**. Countdown from N is completed when $\overline{C}_o$ of all stages are at **0**. At this point the gate (negative-logic NAND) output produces a **1** (for one clock cycle). This is the circuit output ($f_o = f_c/N$) and also the preset pulse which restarts the cycle. Note that the output is not a symmetric square wave. If symmetry is required, the input frequency should be made $2f_c$ and the circuit followed by a standard divide-by-2 counter. Expansion to three or more stages requires only that the C_o of all stages be connected to the gate (three or more inputs).

Digital frequency multiplication is more difficult than frequency division. A $\times 2$ frequency multiplier can be made by differentiating the input wave and then counting both positive- and negative-going edges. In Fig. 16-16a, the NOR gate receives a positive input pulse from one input or the other each time the input makes a transition. Thus the number of pulses out is twice that of the input. The output, of course, is not a square wave. Because sufficient time must be allowed for the RC decay, the circuit speed is more limited than the gate speed.

A more generally applicable but more elaborate method of frequency multiplication utilizes a PLL with a frequency divider (divide-by-N) inserted between the phase detector and VCO. In many IC versions of

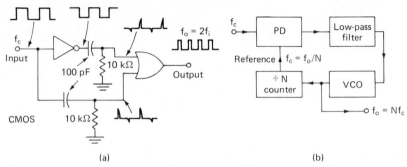

(a)   (b)

Figure 16-16 Frequency multipliers: (*a*) edge counter (×2); (*b*) block diagram of PLL method.

the PLL, for example, the 565, the amount of the VCO is a square wave directly compatible with digital (CMOS) counters, and the connection between the VCO and phase detector is external and therefore separable. This allows a digital frequency divider (divide-by-N) to be added easily. The VCO center frequency is adjusted to approximately Nf_c, where f_c is the input (clock) frequency, assumed symmetric. The output is also symmetric at N times the frequency. A disadvantage of this method is the limited tracking range of the VCO, which limits the output frequency range (unless the VCO center frequency is readjusted). As with any PLL, stability can be a problem.

16-8   Digital Interval Timers

An interval timer measures the time interval between a start and stop pulse with the result displayed on a digital readout (Fig. 16-17). It consists of a digital counter plus a clock (often a crystal-controlled

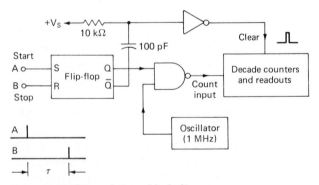

Figure 16-17 Interval-timer block diagram.

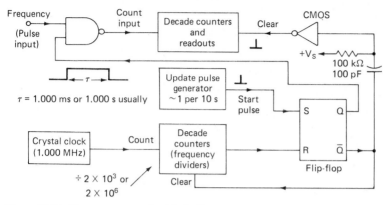

Figure 16-18 Frequency-counter block diagram.

oscillator). Decade counters and crystal oscillators are described else-where.

A 1-MHz oscillator will result in a count equal to the time interval in microseconds. To convert into milliseconds or seconds only requires a shift of decimal point on the counters. If high resolution is not required, the displays can be omitted on the least-significant digits. An alternative approach is to divide the oscillator output by some power of 10. The factor can be selected by a switch which acts as a range control. Often the decimal point is selected by the same switch. If the input-pulse rise times are slow, comparators can be added to the inputs.

16-9   Frequency Counters

A frequency counter records the number of cycles during a fixed time interval, commonly 1 s. As indicated in Fig. 16-18, it is a variation of the decade counter and interval timer (Fig. 16-17). The NAND (or AND) gate allows pulses to pass to the counter during the desired time interval T which is made precisely 1 s on a start-pulse generator. Simultaneously the time-base counters connected to an oscillator (f_o = 1 MHz) are started. When the time-base counters reach N_0 pulses (10^6 in this example), the carry output turns off the flip-flop at a time T equal to $N_0 f_o$, for example, 1 s. During this interval the input pulses (frequency) are counted so that the number of pulses N_f registered is equal to Tf_o. A brief clear pulse is generated as the gates are switched on. Usually the start-pulse generator rate is chosen to be variable in the 1- to 10-s range. Note that the clock frequency dividers are inhibited (held in clear) before the start pulse. Frequencies up to at least 1 MHz with CMOS, 30 MHz with LS-TTL or H-CMOS, and 200 MHz with ECL devices are achievable.

16-10   Decoders and Encoders

The decoders described here have N output lines, only one of which is at 1 at any time, as determined by an input binary (or BCD) number. A 2- to 4-line decoder has four output lines controlled by two input lines, as indicated by Fig. 16-19a. In this example, the decoder is built from individual gates although integrated units are available. As seen from the truth table, an output goes to 1 if the corresponding binary number appears at the inputs; e.g., output 2 will turn on if the input is 01. The extension to 3- to 8-line, 4- to 16-line, 5- to 32-line, etc., decoders is straightforward. A 4- to 10-line decoder (for BCD data) is the same as a 4- to 16-line decoder except that the upper six outputs are missing (Fig. 16-19b). Many other types of decoders are also available.

Some decoders have an enable input capable of turning off all the outputs, regardless of the data input. Note that the D input of the 4- to 10-line decoder acts as an (inverted) enable input if only eight output lines are used (3- to 8-line decoder).

One application of decoders is to minimize the number of lines or wires needed to control or select data; e.g., 4 input lines (bits) can control 16 output lines. Another application is the control or scan of output lines by counters, as described below.

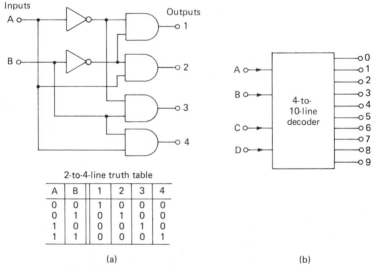

2-to-4-line truth table

A	B	1	2	3	4
0	0	1	0	0	0
0	1	0	1	0	0
1	0	0	0	1	0
1	1	0	0	0	1

(a)   (b)

Figure 16-19 Decoders: (a) 2- to 4-line; (b) 2- to 10-line.

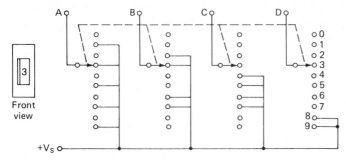

Figure 16-20 Thumbwheel-switch encoder: a ganged 10-position BCD-coded switch is shown. Pull-up or -down resistors are required.

A common kind of encoder is the thumbwheel, a special panel-mounted switch used to enter digital data into an instrument. Perhaps the most popular is the 10-position BCD-coded switch described in Fig. 16-20. There are four output lines corresponding to the BCD code and an input line corresponding to the 0 or 1 voltages. Logic 1 voltage is connected to the common line, and external resistors to ground provide 0 when the switch is open.

16-11 Pulse Sequencing

Often there is a need to generate a series of pulses in a particular time sequence. Techniques can be divided into synchronous and asynchronous methods. Synchronous techniques require a clock control, and all output-pulse transitions occur as the clock pulse makes a transition. Generally some type of decoder is used to generate the output pulse, as illustrated in Fig. 16-21. In this example, two output pulses are produced in sequence, one twice the width of the other. Complex sequences can be obtained by proper decoder design, but the output pulse widths are limited to multiples of the clock period.

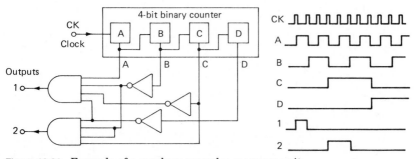

Figure 16-21 Example of a synchronous-pulse sequence unit.

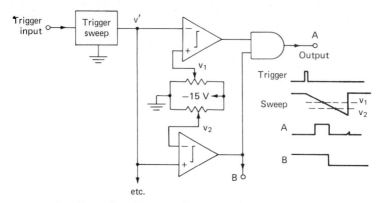

Figure 16-22 Examples of an asynchronous-pulse sequence unit.

A cyclic or repetitive sequence of pulses is produced by the circuit of Fig. 16-21. If a single series of pulse is required, a NAND gate controlled by an *SR* flip-flop can be added (Fig. 16-17) which is switched on by a start pulse and turned off by an output pulse from the counter. Asynchronous pulse-sequencing units can be based on one-shot multivibrators and/or triggered sweeps. An example of the use of multivibrators for a two-pulse sequence is given in Fig. 16-7 and the extension to multiple pulses is straightforward. An example of the triggered-sweep method is illustrated in Fig. 16-22 and is a variation of that shown in Fig. 16-8. Two comparators are used for each output pulse. Both the beginning and the end of the pulse is adjusted by the threshold controls.

Multiplexing

Multiplexing is a technique of transferring several independent signals through a common line or circuit. In time multiplexing, considered here, the several input signals are sampled, transmitted sequentially through a common circuit, and separated at the output or receiver. Presumably the cyclic scanning or sampling rate is fast compared with the rate of change of the signals, so that transmission of sampled portions of the signals is nearly equivalent to transmission of the entire continuous signal. Both digital and analog can be multiplexed.

There are several reasons for multiplexing signals or data. Transmission of two or more signals to a remote site along a single telephone line or radio-telemetry link may be required. In digital systems, especially, data may be multiplexed to avoid duplication of decoders or more complicated devices. Often data are multiplexed simply to reduce the number of signal lines or pins which must be passed in or out of an integrated or printed circuit. The reduction in costs or wiring time can be significant and more than offsets the additional complexity of the multiplexing and demultiplexing circuits at the input and output.

17-1 Analog Switches

Analog (and some digital) multiplexing requires an electronic switch. A convenient switch is the CMOS transmission gate (Fig. 17-1). The basic operation of the gate has already been described. If V_{SS} is grounded, the signals must always be positive, as is normally the case with digital signals. If analog signals of both polarities are to be switched, a dual power supply is needed, for example, $V_{DD} = +5$, V_{SS} or $V_{EE} = -5$ V. The signals for the 4016 must lie between V_{DD} and V_{SS} and the control signals must likewise span V_{DD} and V_{SS}, an inconve-

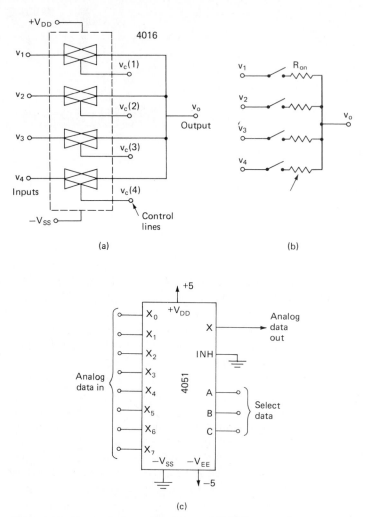

Figure 17-1 Transmission gates: (*a*) quad CMOS gates; (*b*) equivalent circuits; (*c*) four-input multiplexer.

nience since usually the digital signals are positive (0 to V_{DD}). Separate interface drivers are not required for many other analog switches such as the 4051 (Fig. 17-1*c*) which separates the analog negative supply ($-V_{EE}$) from the digital (V_{SS}).

17-2 Analog Multiplexers and Demultiplexers

An analog multiplexer consists of a set of analog switches which are turned on in sequence by a counter and decoder (Fig. 17-2). The

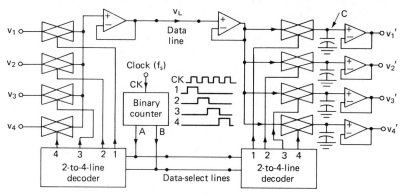

Figure 17-2 Four-channel multiplexer and demultiplexer.

composite signal is transmitted along the lines (Fig. 17-3). Signal amplifiers and line drivers may be required. Analog multiplexers are frequently employed at the inputs to A/D converters so that a single converter can serve multiple analog inputs.

A demultiplexer at the receiver end is identical to the multiplexer. Capacitors hold the voltage for the time between samples in the version shown. This method is similar to that employed in the track-and-hold circuit (Sec. 17-7). The output follows the respective input signals in a stepwise form (Fig. 17-3).

Presumably the sampling rate is rapid compared with the rate of change of the signals. According to the Nyquist sampling theorem, the

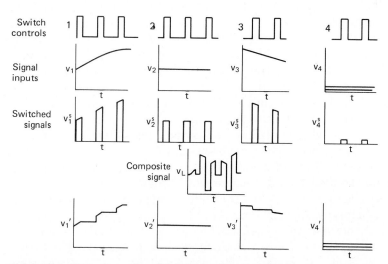

Figure 17-3 Four-channel multiplexer analog signal examples. Normally the signal changes more slowly than shown.

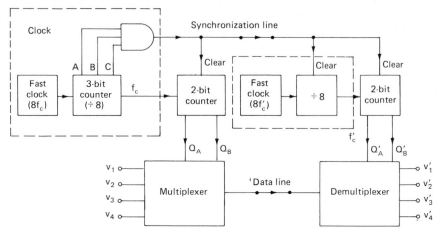

Figure 17-4 Data-scanning clock and synchronization method. Expansion from 4 to 16 on channels is straightforward.

sampling frequency f_s must be at least twice the highest signal frequency. It need only be slightly higher than the signal frequency if a sharp-cutoff filter (high-order Butterworth) is used at the output. With the simple capacitor filter shown, however, f_s must be one or two orders of magnitude higher.

Synchronization of the multiplexing counters and decoders can be done most simply by driving the respective decoders (data selectors) by the same counter, as shown in Fig. 17-2. Extra lines can be avoided by utilizing separate clocks with almost identical frequencies (f_c and f_c') at the transmitter and receiver ends (Fig. 17-4). Some means of synchronization is necessary however. In this example there is little point in multiplexing because the number of lines running from the multiplexing side (transmitter) to the demultiplexer side (receiver) is only reduced from 4 to 3, but with a larger number of channels there is a greater reduction in the number of lines required. It can be seen that it would be desirable to reduce the number of data-selection lines as is done in Fig. 17-4, where only one line (synchonizer line) in addition to the data line is required. The frequencies of the two independent clocks are approximately but not exactly the same without synchronization. Reset of the demultiplexer or receiver clock when the multiplexer clock reaches 0 is accomplished by the synchronization pulse. To avoid extreme shortening (or lengthening) of the demultiplexer clock pulse, which would occur if the synchronizer line or reset pulse happens to occur in the wrong phase of the clock cycle, a fast clock ($8f_c$) with a divide-by-8 counter is employed. Reset occurs when the 3-bit counters reach 0 and the uncertainty is only one-eighth clock cycle (or one fast clock cycle). Without the extra divide-by-8 counter, the uncertainty in reset time is nearly one clock cycle, and thus the demultiplexer-counter Q_A output on count zero could range from a tiny fraction of a clock cycle to nearly two clock cycles long, an unacceptable situation. The extra divide-by-8 counter may be thought of as providing a means of resetting or synchronizing the phase of the receiver clock. Another advantage of the counter is that subcycle timing, if needed, can be controlled through a decoder. This synchronization method is widely used in serial data transmission, especially when it involves UARTs (see Sec. 19-3), except that the fast clock is $16f_c$.

17-3 Latches

A latch is a set of flip-flops, usually four, six, or eight to a package, used to store binary data temporarily. Often they are connected to counter (BIN or BCD) outputs to hold data for display while the counters are cleared and recycled. Both SR and type-D flip-flops are employed. When the latch enable of the SR-type flip-flops (Fig. 17-5a) is high, the latch flip-flops follow the inputs and therefore the input and outputs are identical. When the enable goes low, the flip-flops hold the data and therefore the outputs are equal to the input data when the enable is switched.

Latches built from type-D flip-flops (Fig. 17-5b) transfer data from the input to the output on the rising edge of the clock (usually) or

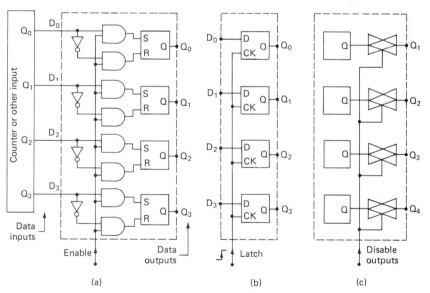

Figure 17-5 Examples of 4-bit latches: (a) $S + R$ type; (b) type D (clocked); and (c) tristate output.

latch-enable pulse. These differ from the RS type in that the output does not change if the input data change while the latch enable remains high.

For multiplexing, a tristate latch is desirable. It consists of a standard latch with a transmission gate connected to the output. The name tristate implies that a third high-impedance (or disabled or disconnected) state exists in addition to **1** and **0**. This disconnect feature allows the outputs of two or more latches to be connected in parallel. Also it is convenient where a single line is used for data input and output. In operation, care must be taken to see that the outputs of all but one of the parallel latches are disabled to avoid shorting outputs.

17-4 Digital Switches

Digital data may be switched by transmission gates or by combinations of gates (decoders). The four-channel data switch shown in Fig. 17-6a is essentially the same as the analog switch described above (but with V_{SS} grounded). If the decoder has an enable E input, all the gates can be turned off. In this case the output is tristate. In Fig. 17-6b, the appropriate data line is switched on by an AND gate, controlled by a decoder. Expansion to more than four channels is straightforward.

The OR gate can be replaced by an open-collector or wired-OR gate. When all lines are turned off (all AND gate outputs at **0**), the output is in the off state. However an external pull-up resistor is needed. Other similar devices can be tied to the same output and turned on when

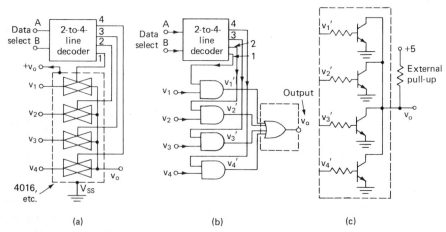

Figure 17-6 Digital data switches: (a) CMOS transmission-gate type (see Fig. 17-1c); (b) AND-OR-gate type; (c) open-collector (wired-OR) option.

required. In this sense the wired-OR acts as a tristate output (Fig. 17-6c).

17-5   Display Multiplexers

Multiplexing of seven-segment LED displays with a number of digits is done primarily to reduce the number of decoders and drivers required. Each digit is turned on or scanned in sequence at a fast rate (>100 Hz), so that the flicker cannot be discerned by the eye. The scanning technique is similar to that for the analog multiplexer except that the four BCD data lines are switched simultaneously. Only one BCD-to-seven-segment decoder is required. All the seven-segment LED (Fig. 17-7) common-anode displays are connected in parallel, but only when the anode of a particular display is connected to the positive supply by the driver transistor will that display light. Calculation of the current-limiting resistance R_L is the same as for a one-display segment except that it is chosen such that the average rather than the peak current is equal to the dc value per segment. Switching of the anode driver and latch output is synchronized by the decoder-data selector driven in turn by the scan clock and counter. Typically the scan frequency, which is not critical and need not be synchronized with any other operation, is 0.1 to 10 kHz.

Multiplexing an LCD is more difficult than an LED display because

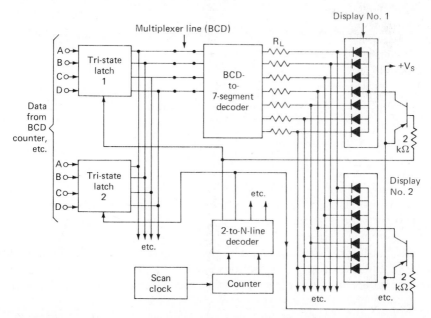

Figure 17-7 Seven-segment LED display multiplexer. Only two channels are shown for simplicity.

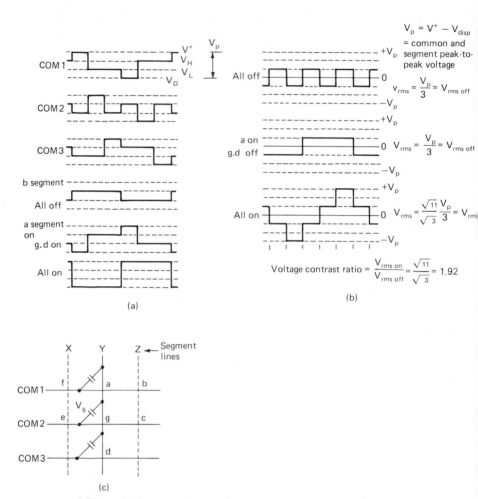

Figure 17-8 LCD multiplexing scheme: (*a*) contrast-voltage characteristics; (*b*) block diagram; (*c*) row and column drive voltage waveform.

the segments or dots do not turn off by reverse polarity. Indeed they must be ac-excited, that is, the average dc component of the drive (composite square wave) must be zero. However, LCD segments do not turn on unless a threshold voltage (about half the saturation voltage) is reached (Fig. 17-8*a*). Thus a multilevel (or backplane and segment) drive is capable of switching a pixel (dot or segment) on or off if the voltage difference across the electrodes at a particular time is above or below threshold. Often a three-level or one-third bias drive is used. For simplicity a 3-V supply along with an 8:1 duty cycle is assumed in Fig. 17-8*c*. Here the pixel is turned on at clock cycle **0** by pulsing it first to full positive ($+3$) and then to full negative (-3). At other times it is off (±1 V). Note that the average (dc bias) voltage is zero. Specialized decoder/drivers such as the ICM 7231 are required.

17-6   Serial-to-Parallel Digital Data Conversion

Digital data are normally presented in parallel form, i. e., as the states (**0** or **1**) of a set of flip-flops or a set of lines equal to the number of bits of data. All the data bits are available simultaneously. In serial form the data are transmitted sequentially, 1 bit at a time along a single line. One data bit is transmitted per clock pulse.

Conversion to, and transfer of, serial data usually is implemented with D flip-flops, as illustrated in Fig. 17-9 for 2 bits (more bits simply require additional flip-flops). According to the truth table for the D flip-flop (Chap. 3), the state present at the D input before the clock transition will appear at the output Q after the clock transition. Since the output of one flip-flop is the input of the next, a particular state assumed preloaded (say **1**) will be transmitted serially from one flip-flop to the next. To move the bit by n flip-flops, n clock cycles are required. Transfer of serial data of a standard 8-bit register (set of flip-flops) therefore requires exactly eight clock cycles (additional clock cycles would transmit zeros or meaningless data).

Conversion of serial data back into parallel form is done by a similar set of D flip-flops except that the data are transmitted into the inputs of the first D flip-flop, as shown. After eight clock cycles for an 8-bit register, the data originally loaded into the transmitting register will have been transmitted to the receiving register, where it is available in parallel form (Fig. 17-10).

Loading of the original data (parallel form) into the transmitting register is done before the first clock cycle. Since some means of

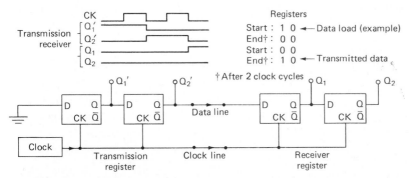

Figure 17-9   Serial-output D flip-flops (2 bits). For simplicity, loading and synchronization techniques are not shown.

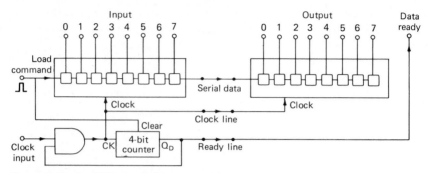

Figure 17-10 An 8-bit parallel-input–serial-output and serial-input–parallel-output register connection.

synchronization must be provided, the data are taken from the receiving register only after the required number of clock cycles following loading of data into the transmitting register. Obviously, if the data are taken out even one cycle too soon or too late, the data are entirely erroneous.

Elimination of the clock line is possible by transmission of synchronizing pulses along the data line. Sometimes a higher-voltage synchronization pulse (if allowable) is sent and detected at the other end by a comparator set to reject the standard-level signals. The pulse synchronizes a clock at the receiver end, as described in connection with the analog multiplexing technique above. A more common synchronization technique is to add similar synchronization (start and stop) pulses to the data stream. Separation of the data and synchronizing pulses requires a substantial amount of additional logic or a special-purpose IC. The complete serial transmitter-receiver unit, which includes clocks and synchronization logic, is referred to as a UART and is described in the next chapter.

In Fig. 17-11 CMOS series-parallel registers are utilized for serial data transmission. Parallel input (to the 4021) occurs if the parallel-

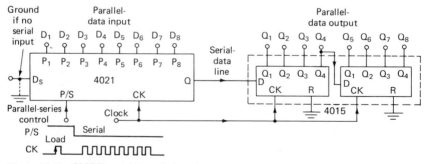

Figure 17-11 CMOS serial-parallel register.

series control is high. Data are clocked out (serially transmitted) on the rising edge of the clock pulses if parallel-series control is low. After eight clock pulses the data are positioned in the receiver register (4015). A counter (3-bit, not shown) must keep track of the clock pulses and thus control the transmission process.

17-7 Sample and Hold

A sample-and-hold or track-and-hold circuit is a switched analog amplifier with an output which is equal to the input when switched to the track mode but keeps the output constant at its current value when switched to the hold mode (Fig. 17-12a). It consists of a FET switch or transmission gate followed by a capacitor. When the gate is on, the capacitor tracks the input. A low-output-impedance (unity-gain) am-

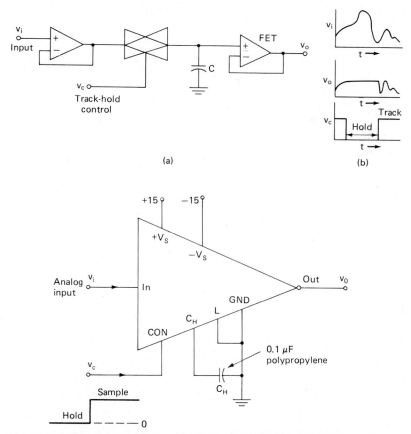

Figure 17-12   Track and hold: (a) circuit; (b) typical response; (c) chip version.

plifier supplies enough current to charge and discharge the capacitor rapidly ($\tau = CR_{on}$). A high-input-impedance (FET) op-amp (also unity-gain) minimizes capacitor discharging in the hold mode.

Analog multiplexers often employ sample-and-hold circuits on the inputs, especially when connected to A/D converters. The voltage is held during conversion to eliminate errors which might result from rapidly changing signals.

Single-chip sample-and-hold ICs (Fig. 17-12b) not only are simple to use but usually respond faster (10 μs acquisition time). Another advantage is that the stray capacitance between the control line and capacitor is generally smaller so that the step in output voltage as a result of switching is small (<1 mV).

17-8   Charge-Coupled Devices

Charge-coupled devices (CCD), the related bucket-brigade devices (BBD), as well as switching capacitor filters (Chap. 10) are special cases of ICs termed charge-transfer devices (CTD). They are capable of storing and transferring charge between capacitance-storage elements under clock control. Input signals are subdivided into a series of time segments for storage, as in time multiplexing. Both analog and digital signals can be handled.

Basically these devices are a series of linearly connected FETs on a single substrate (Fig. 17-13). The gate-channel capacitance stores the charge. Transfer from one FET to another requires a sequence of voltages to be applied to the gates as synchronized by the multiphase clock. Initially the charge is injected into the input capacitance in proportion to the input signal. The ac signal is superimposed on a positive dc bias. Next a gate-voltage bias is applied such that the input gate (gate 0) is cut off, thus trapping the charge on the first FET stage.

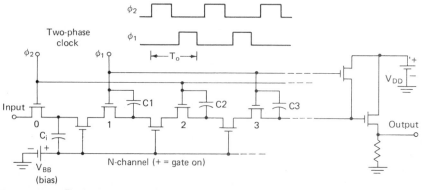

Figure 17-13   Basic charge-transfer-device organization.

Gate 1 is closed, so that at this time no charge flows into section 2, but during the next phase gate 1 is opened by the application of an appropriate control voltage (clock $\emptyset 1$). The charge is then withdrawn by lowering the potential of the channel of FET section 2 (clock $\emptyset 2$). Each complete clock cycle results in transfer of charge from one FET storage element to the next. At the end of N elements (perhaps 64 or 256), the charge is monitored by an output FET, which acts as a unity-gain voltage amplifier. A total of N complete cycles, each of period T_c, is required to transfer the charge from the input to the output. Thus the total delay T_D is NT_c.

A limitation of the CTD is charge loss during storage and transfer. Typically one-ten-thousandth of the charge is lost per transfer and one-ten-thousandth of the charge will leak per millisecond. These losses limit the total time delay. Often the output is amplified to make up the charge-transfer (voltage) loss. Recirculation of the analog signal from the output to the input is possible, but since the voltage transfer even after amplification is not exactly unity, an analog signal cannot be stored for long. Digital signals can be stored indefinitely, however, since they can be restored to standard digital level before being transferred back to the input.

Digital CTDs are also utilized as serial memories. Because of the complexity of the read, write, and data-recirculation logic, these circuits will not be described here. It should be pointed out that serial registers are functionally similar, and the main advantage of the CTD is low cost when a large number of bits are to be stored.

18

Analog Data Transmission and Recording

Most direct analog data transmission links and recording techniques are restricted to ac, typically in the audio-frequency range (20 Hz to 20 kHz), implying that dc signals cannot be recorded directly. With the additional problem of gain variation, which is difficult to eliminate in both direct transmission and recording, some form of modulation or encoding is needed to transmit or record analog data precisely. Here analog methods are described; digital methods are deferred to the next chapter.

18-1 FM Analog Data Transmission

Frequency modulation and demodulation techniques are discussed in Chap. 14. The phase-locked loop is particularly suitable for accurate FM demodulation. With frequency modulation (Fig. 18-1), a difference in frequency Δf with respect to a center or carrier frequency f_o is proportional to the signal or modulation voltage v_m

$$\Delta f = f - f_o = K_m v_m \tag{18-1}$$

where K_m is a modulation index. There is no lower frequency limit on V_m; that is, dc can be transmitted (dc corresponds to a constant Δf). The carrier frequency is set at the center of the transmission band; for example, $f_o \approx 1.8$ kHz for a frequency of 0.3 to 3.3 kHz. The maximum modulation frequency of a sinusoidal signal component $V_m(t) = V_m \sin 2\pi f_m t$ permissible without distortion is about 10 to 30 percent of the carrier frequency.

Carrier amplitude variation due to a variation in transmission link gain does not affect demodulated signal amplitude because signal

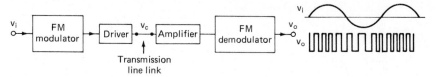

Figure 18-1   Analog FM telemetry block diagram.

amplitude depends on frequency and not carrier amplitude. The price paid for FM encoding and its associated advantages is a reduction in bandwidth to 5 to 30 percent of that possible with direct (ac or audio-frequency) transmission. Also circuit complexity is greater.

A simple RF telemetry transmitter for short-distance (10 to 100 m) communication is easily assembled from widely available components (Fig. 18-2). It utilizes the Colpitts oscillator (Chap. 11) with a voltage-variable capacitor modulator (Chap. 14) tuned to the standard FM broadcast* band. Even an ordinary signal diode will work as a variable capacitor with diminished sensitivity. A standard FM radio receiver can be used to demodulate the high-frequency signal into an audio signal. If the transmission is pulse-modulated, a comparator (see Fig. 18-9b) can be added to recover pulses from an AM signal while a PLL tone decoder (see Fig. 19-2) will recover pulses from an FM signal or tone.

---

* RF emissions are controlled by law, which in the United States is administered by the Federal Communications Commission. Experimenters are unlikely to run afoul of the law if the dc power into the oscillator (or last RF stage) is under 100 mW and the transmission is in the FM band (88 to 100 MHz) at a frequency which does not interfere with local radio broadcasts.

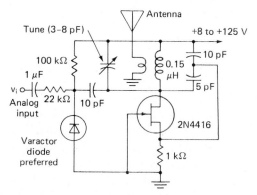

Figure 18-2   Simple high-frequency FM transmitter.

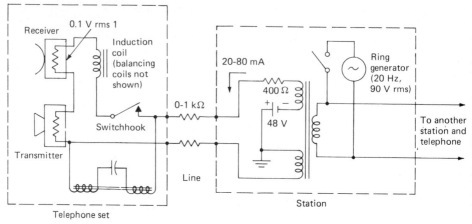

Figure 18-3 Simplified equivalent circuit of a telephone.

18-2 Telephone Transmission Characteristics

Characteristics for telephone transmission were established by Bell Telephone decades ago and newer designs remain electrically compatible.* A dc source (48 V) with a limiting resistance at the station supplies current for the microphone (Fig. 18-3). As the microphone resistance varies at the audio frequency, the current is modulated. The audio signal modified by a transducer or induction coil is transmitted to the headphone at the other instrument. Nominal ac impedance is 600 Ω, but line resistance may increase this several times for longer runs and an extension phone will drop it by 50 percent. When the instrument switch is closed (off hook), the terminal dc voltage drops to about 6 V and the dc current flow (20 mA min), detected by the station, produces the answer mode. A 600-Ω resistor connected (e.g., relay contacts) across the two terminals is electrically equivalent to picking up the telephone receiver. Other characteristics are listed in Table 18-1.

18-3 MODEMS

A MODEM (modulator-demodulator or data set) is an instrument which converts tones on a telephone line into digital-level data and the reverse. It implements the frequency-shift-keying modulation method, which is discussed further in Chap. 19. Many computer (teletype or

---

* The terminal designations of T (tip) and R (ring) refer to the old jack-type switchboards.

TABLE 18-1 Selected Characteristics of U.S. Telephone Line Transmission

Parameters	Typical value	Range
Supply voltage	−48 V dc	−47 to −105
Operating current	40 mA	20 to 120
Loop resistance	800 Ω	0 to 3.6K
Set resistance	200 Ω	100 to 400
Loop loss	8 dB	0 to 17
Ringing signal	20 Hz, 90 V rms	16 to 60 Hz
		(40 to 130 V)

Tone	Frequency, Hz	Time on/off, s
Ring	20	2/2
Dial	350 + 440	Continuous
Busy	480 + 620	0.5/0.5
Ringback	440 + 480	2.4 or 1/3

Dialing row/col. no.	Column frequency, Hz	Row frequency, Hz
1	2944	5104
2	2688	4640
3	2432	4176
4	2176	3828

CRT) terminals require a MODEM as an interface unit. In the transmit mode, serial input data are clocked out at a predetermined rate in the form of audio tones. The tone pulses are connected to the line by a transformer (hard-wired) or through an acoustical coupler, consisting of a microphone and small loudspeaker mounted in a cradle shaped to fit a standard telephone headset. The level of the ac signal at the coupler terminals (input and output on the same line) is roughly the same level as on the telephone line (0.05 V on receive, 0.3 V on transmit).

A MODEM which can transmit and receive at both ends is termed duplex. If one unit can only send and the other only receive, the operation is termed simplex. If both units do not transmit data at the same time, the operation is half-duplex, but if both transmit and receive simultaneously on the same line, the operation is full-duplex. Under full-duplex operation the transmitted and received signals operate at separate frequencies. Normally in full-duplex operation the received signal is retransmitted, or echoed, back to the originating MODEM as a check on transmission accuracy. Even in half-duplex operation the reverse transmission of a steady tone is standard so that the data-transmitting station has a check that the receiver station remains connected. Most MODEMs can operate in either half-or full-duplex modes without modification; the mode is controlled by the connected terminal.

The American (Bell Telephone) frequencies are shown in Table 18-2.

TABLE 18-2　Standard MODEM Frequencies

| Logic | 110 and 300 b/s | | 1200 b/s (half duplex), Hz |
	Transmit, Hz*	Receive, Hz*	
1 (mark)	1270	2225	1200
0 (space)	1070	2025	2200

Frequency tolerance ± 5%.
* For originate MODEM.

Often the old teletype terms "mark" and "space" are used to designate **1** (high) and **0** (low), respectively. The MODEM which initiates the call, termed an originate MODEM, transmits at the lower-frequency band and receives at the higher. The answer MODEM responds by transmitting on the higher-frequency band and receives on the lower. Usually the remote terminal originates the call and the computer answers. Various intervals defining maximum and minimum contact and disconnect times have been established.

A block diagram of an originate MODEM is given in Fig. 18-4. After dialup by a standard telephone handset, the line is connected to the MODEM. The duplexer separates the transmitted and received signals. Its main purpose is to keep the transmitting signal (especially second harmonics) from interfering with the received signal. A notch filter keeps the transmitting frequency and harmonics from the receive section.

The tone demodulator may be a phase-locked loop (low baud rates) or a digital detector based on tone frequency or period. The receiver bandpass filter must be carefully designed to allow sharp discrimination against interfering signals without producing excessive ringing or phase differences between the two received frequencies. During transmission the voltage-controlled oscillator switches between the two transmitting frequencies. Often a standard RS-232 interface is employed, as discussed in Sec. 19-5.

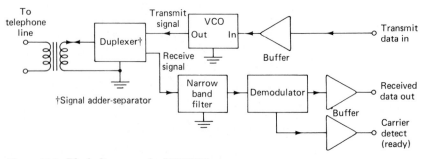

Figure 18-4　Block diagram of a MODEM.

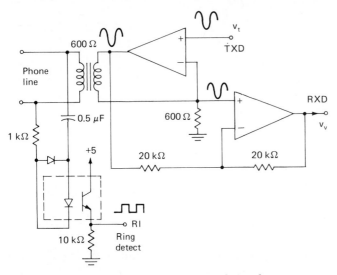

Figure 18-5 Telephone duplexer and isolated ring detector.

A duplexer which separates the transmitted and received signals based on operational amplifiers is shown in Fig. 18-5 (a special multiwinding transformer can also be used). The signal to be transmitted V_t produces a current i_t through the transformer and R_i. If the transformer load is exactly 600 Ω, the voltages across R_i and the transformer will be equal. The difference amplifier which has unity gain on the inverting side but a gain of 2 for the other side subtracts signals resulting in a zero output V_r for the received signal. An external signal applied across the transformer (with $V_t = 0$) will produce an equal voltage at the output V_r. Since the line impedance deviates appreciably from minimum (600 Ω), some signal will be transferred from V_t to V_r and complete separation of the transmitted and received signal is not practical.

Telephone lines are operated balanced and neither side may be grounded (transformer or other isolation is necessary). Optical isolation may be used, for example, to detect the incoming ring (Fig. 18-5).

18-4 Analog Tape Recorders

Magnetic tape recorders, including the home entertainment type, are based on amplitude modulation of the tape at an audio frequency. As discussed above, any device which can record an audio signal can be used for recording analog or digital signals by proper modulation-demodulation methods. It is instructive to discuss the recording and

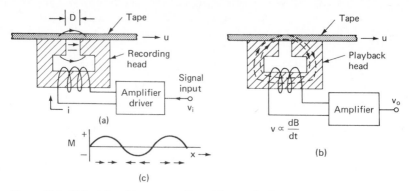

Figure 18-6 Principle of tape recording: (*a*) record; (*b*) playback; (*c*) tape magnetization waveform.

playback principles in order to understand the frequency and dynamic-range limitations.

A magnetic tape has small magnetic particles embedded in the plastic matrix. The particles can be magnetized by an external field produced by the recording head (Fig. 18-6). Current-drive requirements for the recording-head coil are similar to those of loudspeakers. The head has a narrow gap filled with a nonmagnetic material across and close to which the magnetic field is concentrated. The field drops off very rapidly away from the gap and can be neglected elsewhere. All particles within the gap at a given instant have approximately the same magnetization. Assuming that the head-drive current is sinusoidal with time, the magnetization M along the tape will be sinusoidal with distance x, with $x = ut$, where u is the tape velocity. Only a fraction of a wavelength λ, at most a half wave, can be magnetized at any instant. Since the length of the half wave on the tape cannot be shorter than the gap distance D, the highest frequency that can be recorded corresponds to a wavelength of $2u/D$. High-frequency recording therefore requires a high tape speed or a small recording-head gap (1 to 4 μm). Small cassette recorders typically have a frequency limitation of about 4 to 10 kHz, while a reel-to-reel recorder (15 in/s) may respond above 40 kHz.

The playback head has the same general construction as the recording head. During playback the magnetized particles near the gap produce a small magnetic field B in the head. A voltage proportional to dB/dt is induced in the coil. For a sinusoidal wave ($M_o \sin \omega t$), the induced voltage will be proportional to $\omega \cos \omega t$; that is, the output voltage will decrease as the frequency is reduced and will be zero for direct current. Response below 10 to 50 Hz is generally impractical.

Tape is erased by exposing the tape to a high magnetic field. An erase head similar to the recording head may be used, operated at a frequency higher than the maximum response frequency to avoid interference with the recorded signal. A high field magnetizes the particles to the maximum extent, thus wiping out the magnetization due to previous signals. Alternatively the entire tape may be exposed to a strong 60-Hz field, which is slowly reduced by moving the tape away from the coil. As the tape passes from the high- to zero-field region, through slowly diminishing magnetic hysteresis loops, the net magnetization approaches zero. Usually a moderate-amplitude high-frequency component or ac bias is added to the recording-head current. It serves to linearize the current-magnetization response curve, especially for small signals. The bias signal is too high in frequency to be played back. For an audio recorder (20 Hz to 20 kHz) the bias frequency may be 100 kHz.

Fluctuations in tape speed, a common problem with low-quality recorders, results in a frequency variation in playback. Frequency variations are reflected in baseline variations with FM analog recording. It is a problem also with digital tone demodulation if the frequency shifts outside the bandwidth of the receiver, a distinct possibility. Amplitude variations, which might be caused by warped tape or an uneven distribution of magnetic particles in the tape, are always present with audio (AM) recordings. Assuming that the signal does not completely drop out, these variations do not adversely affect most analog and digital modulation methods because they are designed to be insensitive to amplitude. Playback signal waveforms are discussed in the next section.

A standard audio playback circuit (Fig. 18-7d) consists of a high-gain audio amplifier with a filter to compensate for the reduction in pickup-head voltage at lower frequencies. Basically the filter is a low-pass filter with the break frequency set equal to the lowest frequency to be played back, for example, $f_1 = 50$ Hz. All signals are attenuated in proportion to frequency, thus compensating for the linear increase in induced voltage with frequency. A typical playback voltage (for a constant-amplitude-current record) is given in Fig. 18-7a. The fast drop in amplitude at high frequencies is due largely to the record-head wavelength limitation discussed above. A second high-frequency filter is also added to reduce the effect of tape noise (hiss). While tape noise is fairly uniformly spread over the frequency band, it is more audible at higher frequencies. A filter is added to the record circuit to emphasize frequencies above a chosen point ($f_2 = 1.77$ kHz for the National Association of Broadcasters standard at a tape speed of $1\frac{7}{8}$ in/s). During playback a deemphasis filter is added to reduce the gain above the breakpoint. More precisely, during playback the

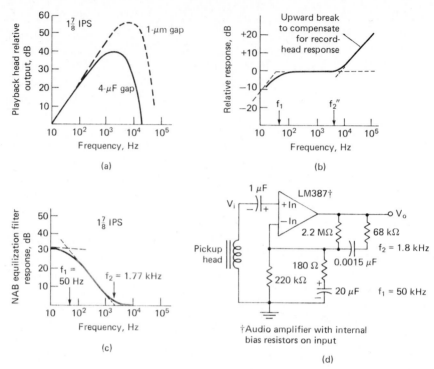

Figure 18-7 Audio-frequency tape recorder: (*a*) uncompensated playback response; (*b*) record characteristic; (*c*) playback characteristics; (*d*) playback amplifier.

compensation filter which reduces the signal in proportion to frequency is set so that it breaks to a horizontal slope above f_2 (Fig. 18-7*c*).

Recording of analog signals, which may have a dc component, may be done by frequency or pulse-width modulation, as discussed below. With the FM method, the carrier frequency f_o is often recorded on a second track to compensate for tape-speed variations, possibly by controlling motor speed.

18-5 Digital Tape Recorders

One technique of recording detailed data is to tone-modulate an analog recorder. This frequency-shift-keying (FSK) technique is similar to that described for modem telephone line communication (Sec. 18-3) except that the frequencies are less precise. Accurate frequency control is not practical with most medium- or low-quality cassette recorders intended for home entertainment because speed control is mediocore. A more suitable method for low-cost recorders is the Kansas City standard. It uses two widely separated frequencies which need not be specified precisely. At a nominal data rate of 300 b/s, **0** consists of 4 cycles of 1200 Hz and **1** is 8 cycles of 2400 Hz. The tones are conveniently generated by frequency division (divided by 2 or 4) from the same clock that drives the transmitter UART (Chap. 19) normally used with the tone encoder. In the receive mode, a 2400-Hz clock can be derived from the recorded data itself, and this drives the receive UART.

A more efficient technique is to record the digital data directly. The tape is magnetized to the saturation level (positive for **1**, negative or reverse saturation for **0**), as indicated in Fig. 18-8. While a dc signal (sustained **0** or **1**) cannot be picked up on playback, the change in logic state can be as a positive or negative pulse. The pulses, often detected by a pair of comparators, trigger an *SR*-type flip-flop to restore the data to standard digital level. No clock or regular data rate is required. There is a restriction on the maximum rate commensurate with the high-frequency limitation of the tape recorder.

Usually data are recorded at the highest reliable rate in order to maximize storage density and minimize data transfer time. Waveforms become distorted at high rates in the sense that the rise time becomes comparable with the bit time and that the zero and compar-

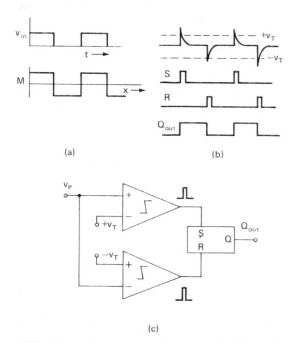

Figure 18-8 Digital recording: (*a*) recorded tape signal; (*b*) playback signal; (*c*) block diagram of digital signal resoration current.

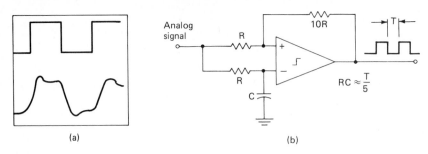

Figure 18-9 Recorder playback waveforms: (*a*) square wave at clock rate; (*b*) comparator circuit.

ator threshold levels become uncertain (Fig. 18-8). If the ratio of high to low pulse widths is limited, the comparator threshold level can be established reliably with a simple averager (Fig. 18-9*b*). Encoding techniques for which the polarity reverses at least once per clock cycle, such as Manchester code, are further discussed in Chapter 19.

Larger computers employ a wide nine-track tape which records 8 data bits and a clock channel at once.

18-6 Magnetic Disk Recording

The principle of recording on a hard or floppy magnetic disk is the same as that of the tape recorder. The embedding magnetic particles and the recording/playback head of the disk and tape are similar. Recordings are organized (Fig. 18-10) into circular tracks or sectors (typically 10 sectors for a $5\frac{1}{4}$-in disk, of which 8 or 9 are used). Track beginnings and ends are marked by a distinctive pattern. These patterns are usually not built into the (soft-sectored) disk but are recorded during an initialization process termed "formatting."

The record/playback head when loaded or engaged rides on a cushion of air close to the disk. A small gap is maintained between the head and disk (and in the head itself) in order to achieve good high-frequency response and high data density. Disk rotation speed is 30 revolutions per second (rps) typically so that one track of one sector can be read in 0.3 ms. Movement of the head radially from one track to another during the seeking process requires a maximum of 50 ms for a $5\frac{1}{4}$-in disk.

18-7 Bubble Memories

Like tape or disk recorders bubble memories store digital data as localized field reversals in a magnetic film. In the thin, single-domain

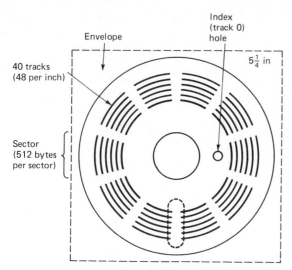

Figure 18-10   Magnetic disk recording track organization.

type of magnetic materials, the magnetic moments of all molecules are initially aligned (by a high magnetic field) in the same direction to form a bubble-free state. If a very localized reverse magnetic field is applied to the magnetic film by passing current through a small wire loop, a small group of molecules becomes magnetically reverse biased. When the local reverse field is removed, a disturbed cylindrical region or bubble persists if a uniform external field of the proper strength is present. The bubble is equivalent to one bit of stored data. A strong forward magnetic field applied to the bubble removes it, thus a bubble can be created or destroyed at a particular spot by electrical impulse. Bubbles can be confined to desired paths by placing thin strips of magnetic and conducting material on the film to produce local energy minimums. The cheveron-shaped pattern (Fig. 18-11) confines the bubbles to a series of specific locations. Bubbles can be moved from one chevron to another by applying a rotating external magnetic field.* This process allows data to be stored and moved serially under control of a clock. A bit advances one position each field rotation or clock cycle. The bubble can be detected as an induced voltage in a pickup loop, a result of the magnetic field change as the bubble passed a readout location.

Data paths are usually organized in long loops. Data are written in at the path start (write or generator/annihilator location) and, after a

---

* H. Chang,"Magnetic Bubble Memory Technology," Marcel Dekker, New York, 1978.

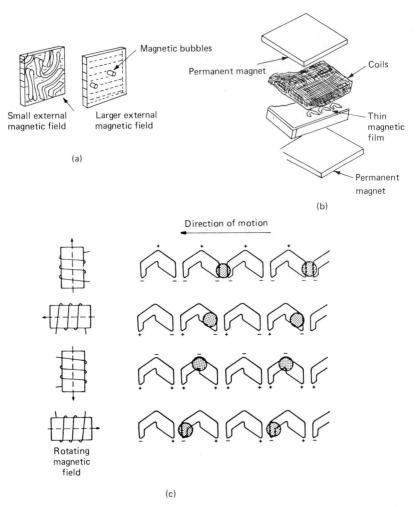

Figure 18-11   Bubble memory: (*a*) bubble formation; (*b*) geometry of memory; (*c*) bubble motion along a chevron path.

certain number of clock cycles, appear at the end (read or detector location). The read and transfer process may involve destroying a bubble so a generator is usually coupled with a detector. If no clock is applied, the data remain in place with power off indefinitely, that is, the memory is nonvolatile. Usually there are two sets of loops, a major loop connected to the input/output devices which function as an internal data bus and a large number (perhaps 100 to 300) of minor loops connected to the major loop which actually stores the data. Each minor loop may hold 256 bits (one page) of data. A typical bubble memory has a data rate of 85K bits/s and can store one megabit.

There are three main parts to a bubble memory system. One is the thin-film bubble memory itself, which is the same size, complexity, and packaging as an IC. It is sandwiched between permanent magnetic and rotating (ac) field driver coils. A complex integrated circuit controls the detection, driving, and sequencing functions. It acts as an interface to the 8-bit host microprocessor.

19

Digital Data Communication

Techniques for transmitting digital data are similar to those employed in transmitting analog data except that only two voltage levels (0 and 1) are required. Modulation or encoding is often necessary since most telemetry links do not transmit dc. Typically only one line of serial data is transmitted. Data in parallel form must first be converted into serial form by a UART. Serial data are formatted both at the bit and byte level. Standards also have been established for the interconnection of data lines to facilitate matching of devices from different manufacturers.

19-1 Frequency Shift Keying and Tone Decoding

Frequency shift keying (FSK) is a type of frequency modulation where digital states 0 and 1 correspond to two different specified frequencies. The MODEM (Sec. 18-3) is an example of an FSK device. In a more typical example, the carrier is at a high frequency (1 to 100 MHz) and the frequency deviation is small (perhaps ±5 kHz).

The FSK decoder consists of a FM demodulator and comparator (Fig. 19-1). A PLL is an excellent demodulator, especially those devices which have an internal comparator (Fig. 19-1b). For higher speed with low error detection, attention must be given to the loop filter and stability. In particular, ringing of the signal at the comparator input should be avoided.

Modulation consists simply of applying the attenuated digital signal to the control input of a voltage-controlled oscillator. Often the VCO section of a PLL is used.

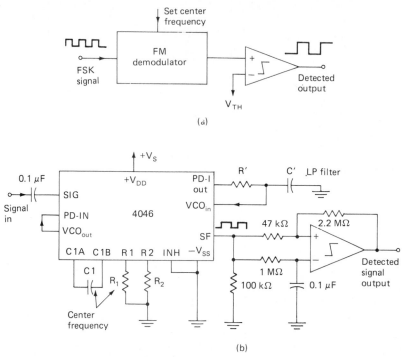

Figure 19-1 FSK demodulator: (*a*) block diagram and (*b*) PLL circuit.

In tone modulation, the ac signal is switched on and off, usually an audio frequency. Detection is usually done by a specialized PLL or tone decoder. The 567 tone decoder (Fig. 19-2) utilizes two phase detectors (PD), one in-phase (I) and one quadrature (II). When a tone of the proper frequency is present and the loop has locked, the phase difference approaches zero and therefore the output of PD-I will be at a minimum. At this point the PD-II out is a maximum. If no tone is present, both will be close to zero. Detector filter time constants are set so that detector I responds more rapidly than detector II. This assures establishment of lock before the comparator connected to detector II responds. Response time is typically longer than 20 carrier cycles and as such is slower than most FSK demodulation.

Often two tones are transmitted together, as is done in telephone dialing, to reduce the chance of noise or spurious tones (speech) being interpreted as a data pulse. Only when both tones, chosen to be harmonically unrelated, are present is the data considered valid, and a pulse appears at the output. Touch-tone dials select one tone by column and another by row (see Table 18-1).

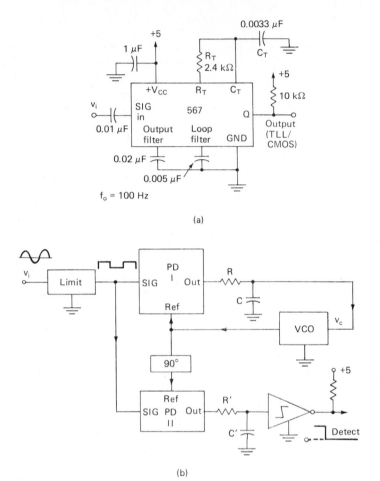

Figure 19-2 A PLL tone decoder: (a) 567 decoder; (b) functional diagram.

19-2 Bit-Level Encoding Formats

Ordinarily digital data are transmitted at a fixed rate which is referenced to a clock square wave (Fig. 19-3). Digital data must be encoded at the bit level into an ac signal for transmission links or recorders that do not pass dc. Bit encoding is not necessary for FSK systems or digital driver/receiver interfaces which pass dc such as the RS-232. For these systems a non-return-to-zero code (NRZ) which has a dc component is acceptable. Indeed the simplest code (NRZ level) is the data itself (or its inverse, Fig. 19-3a). By contrast the transmitted signal of a return-to-zero code (RZ) always goes back to zero during

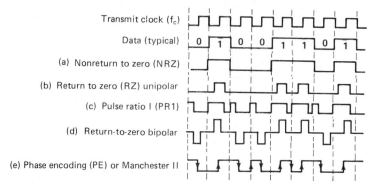

Transmit clock (f_c)

Data (typical)

(a) Nonreturn to zero (NRZ)

(b) Return to zero (RZ) unipolar

(c) Pulse ratio I (PR1)

(d) Return-to-zero bipolar

(e) Phase encoding (PE) or Manchester II

Figure 19-3 Bit encoding format.

each bit. Unipolar RZ codes consist of positive pulses which return to zero for part of each clock cycle. For **0** either no pulse is sent (Fig. 19-3*b*) or a short pulse is sent (Fig. 19-3*c*). Bipolar RZ codes consist of a positive pulse for logic **1** and a negative pulse for logic **0** (Fig. 19-3*d*). The clock itself cannot be reliably recovered at the receiving end with the RZ format of Fig. 19-3*b* since a long string of **0**'s will result in no signal. With the pulse-ratio method where a pulse is transmitted of each clock cycle ($\frac{1}{4}$ clock period for **0**, $\frac{3}{4}$ for **1**) will recover the clock (Fig. 19-3*c*). Note that the data rising edge occurs at the start of each clock cycle.

Phase encoding (Manchester II), a type of NRZ code, utilizes the transition direction of the transmitted pulse at the center of the clock period (rising edge, or data time) to indicate the logic sense. Logic **0** and **1** correspond to the falling and rising transitions, respectively. Data transitions at the beginning of the clock period (phase time) are ignored during decoding and are necessary whenever the serial data do not change sign (**0** followed by **0** or **1** by **1**). With one method of decoding, a one-shot multivibrator triggered by both transmitted data pulse edges (at data time) is used to disable an edge-sensitive latch during the phase time. Advantages of phase encoding are that it has no dc component, it has modest bandwidth requirements, and it is self-clocking. The main disadvantages are the more complex circuitry required and the lack of an inherent error-detection capability. A related format, Manchester I, is similar except that the transition times are shifted by a half clock cycle.

19-3 UART and USART

A universal asynchronous receiver and transmitter (UART) is a relatively complex IC consisting of an independent transmitter and

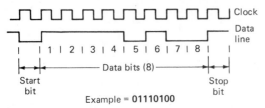

Figure 19-4   Standard UART serial data format.

receiver sections. The transmitter can encode parallel-input data for serial data transmission along a single channel. The receiver performs the inverse function. One UART transmits while another receives. Basically the transmitter is a serial-output parallel-input register (Chap. 17), but additional synchronization information (start and stop bits) is added to the data stream so that an additional (clock) channel is not needed for this purpose. Operation is asynchronous in the sense that the data can be transmitted at any time and need not be at a regular rate.

The clocks of the transmitter and receiver are nominally, but not exactly, the same. Rates of data transmission are chosen through adjustment of the clock rate to be compatible with the frequency response of the line. Standard data transmission rates, among others, are 300 and 1200 bits per second (b/s, or baud).

The synchronization method involving an internal counter requires further discussion. A common format, as illustrated in Fig. 19-4, has 10 bits per word, 8 of which are data bits. There is one start bit, which is **0**, and one stop bit, which is **1**. Other variations such as two stop bits and addition of a parity bit (ninth data bit) are pin-selectable options. The first pulse phase synchronizes the receiver clock by clearing on the leading edge. Only 8 bits appear at the data output bus since the start and stop bits are only used internally. The start bit can either follow immediately after the stop bit or after a longer period of time, as would occur with a normal break and resumption of transmission. Only the middle section of each pulse is checked for valid logic levels, since the leading and trailing edges of each bit normally jitter somewhat.

Commonly a UART and MODEM are employed in tandem. The frequency-modulated information on a telephone line is converted into serial digital form by the MODEM and then into parallel form by the UART. One byte (8 bits) is transmitted at a time. When the byte has been received, a ready (receive-buffer-full) status signal is generated (Fig. 19-5) for handshaking purposes. While the byte may represent any binary number, it usually represents one alphanumeric character in ASCII code (see Sec. 19-8). One end of the transmission link may be a keyboard and cathode ray tube (CRT) display while the other might

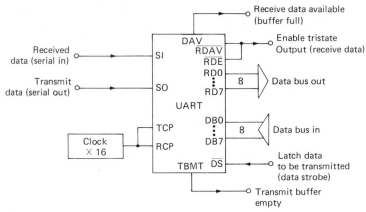

Figure 19-5 Input-output connections to a UART.

be a microprocessor or computer. All UARTs are microprocessor-bus-compatible and can be considered a microprocessor peripheral device (Chap. 20).

The clock input frequency to a UART is a multiple (usually 16) of the bit transmit frequency. An internal binary counter divides the input and allows subdivision of the received bit sequence into several steps. For example, the data check at the middle of the received bit (8 input clock cycles). A parity bit is inserted into the transmit stream as an option. For even parity the sum of all data bits plus the parity bit is even. A flag (output pin) on the receiver is set if a parity error is detected.

A similar device, the universal synchronous/asynchronous receiver and transmitter (USART) has, in addition to the asynchronous UART function described above, a synchronous capability. Data are transmitted continuously in fixed groups of bits, for example, 5 data bits plus a parity bit (odd). Still parallel data input and output are by the byte. Each group (character) has at least one falling edge and this synchronizes the receiver clock. At the beginning of transmission, or whenever no input data are available, a sync character is sent. While the synchronous mode is only slightly faster, it has the advantage of better error-checking capacity.

19-4   RS-232 and Other Interfaces

For short runs standard CMOS-TTL digital lines carrying serial data can be connected directly, but for longer runs (over 1 to 10 m) noise injection or signal degradation may become problems. The RS-232 interface (Fig. 19-6) improves the signal-to-noise ratio by raising the

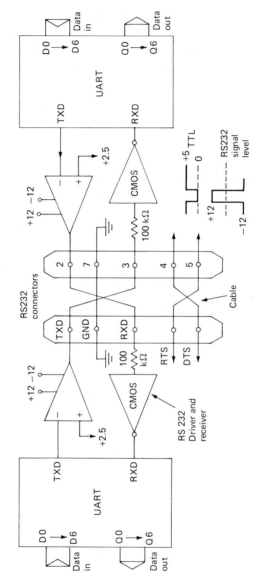

Figure 19-6 RS-232 interface: (*a*) signal line organization (null-MODEM connection shown); (*b*) driver and reciever using standard devices.

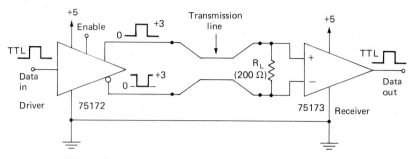

Figure 19-7 RS-422 interface.

signal level to about ± 10 V. The RS-232 is a standard interface for which the characteristics and configurations have been specified in detail. Connector (type-D), signal functions, pinouts, and labels are specified. A special transmitter (or driver) and receiver (a specialized comparator or Schmitt trigger) converts from and to TTL level. An ordinary op-amp and CMOS inverter with an input register is an acceptable substitute if some details such as slew rate are relaxed.

A minimal interface connection consists of only the two data lines and ground (out of the 25 pins), but often two handshaking lines are added. One is data-terminal-ready (DTR), which becomes positive when the remote or originate end is ready to receive data (see MODEMs, Sec. 18-3.) At the computer or answer end, the connections are reversed (the null-MODEM configuration), since data transmitted at one end are received at the other. However, the cable between a terminal and MODEM does not change pin numbers (standard connection) since the MODEM transmit and receive process interchanges the data. Often the RS-232 interface is used as a data link other than between computer and terminal through a MODEM pair, and in this case no convention exists as to whether the standard or null-MODEM connection is to be used.*

Higher speed and higher noise immunity are provided by the RS-422 interface (Fig. 19-7). Data are transmitted over a balanced line at standard LS-TTL level (once side inverted). The receiver is a specialized comparator with a wide input common-mode range so that it can tolerate large noise pulses. Noise often is caused by induced voltage between separated grounds. Noise pulses on the two data lines thus are of nearly equal amplitude and the data difference voltage at the receiver end is little affected by noise. For this reason the differential

---

* Fertile ground for the application of Murphy's law.

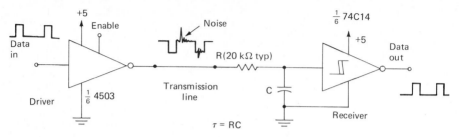

Figure 19-8   Simple CMOS low-power interface.

interface is more noise-immune than the higher-voltage single-ended RS-232 even though the signal voltage level is five times lower.

The RS-422 has the advantage of requiring only one 5-V supply rather than a triple supply needed by the RS-232. Also the driver output can be disabled, thus allowing data transmission in either direction along the same signal pair. However, data cannot be transmitted in both directions at the same time (simplex transmission), unlike most MODEMs which can send in both directions simultaneously (duplex transmission). At high data rates, the reflection of the narrow pulses from long, improperly terminated transmission lines (R_L missing) may cause error.

For short lines and lower data rates (Fig. 19-8), standard CMOS devices may be adequate. Noise has predominantly high frequency components so that a simple low-pass filter added to the receiver is generally effective in noise removal. Of course the data bit or pulse period must be comfortably longer than the filter time constant $\tau = RC$. A Schmitt trigger restores the signal amplitude and edges to normal.

19-5   Fiber-Optic Transmission

Electrical interference can be eliminated entirely by signal transmission over on optical fiber. A fiber-optic cable is a thin transparent glass or (sometimes) plastic fiber through which light can travel for long distances with minimal loss. Around the glass core is a cladding with lower index of refraction (Fig. 19-9). Around this is an air gap and then a protective sheath. Light entering at an angle less than a critical angle (θ_c) will be multiply internally reflected at the core-cladding interface. The acceptance angle increases as the difference between the core and cladding indexes of refraction increases. The cladding eliminates reflection loss which would be caused by surface scratches and dust. Typical glass fibers are 100 to 150 μm in diameter, although

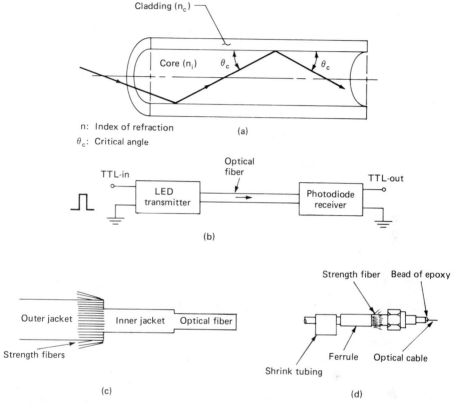

Figure 19-9 Fiber-optic cable: (*a*) geometry; (*b*) block diagram; (*c*) cable construction; and (*d*) connector (also see Fig. 7-20).

single-mode fibers under 20 μm diameter are available, as are plastic fibers 1 mm (1000 μm) in diameter. Connectors have been developed to simplify field splicing of fibers (Fig. 19-9*d*).

Electrically the fiber-optic transmitter (LED) and receiver (photodiode and amplifier) is similar to the optical isolater (see Fig. 5-19 and 5-20). Almost always digital data are transmitted and both input and output are TTL-compatible. Typical data transfer rates are 1 to 10 MHz.

19-6   Bar Code Readers

Manufactured products have attached printed labels with a bar code for the purposes of inventory control and pricing. A reflecting optical pickup is passed across or scans the label, producing a serial digital signal at the output. Sometimes a laser beam (light spot) is scanned

Figure 19-10 Optical
bar codes.

instead and then the reflected signal need only be imaged onto a
photodetector to produce the time-varying signal.

The digital signal is serial like that of a UART, but because the scan
rate normally is not controlled, the bit rate is not fixed. No single
standard format or code has yet evolved, but all are self-clocking (or
clockless) and have a heading or sync pattern at the beginning to
provide reference timing for the code which follows. A typical code is
shown in Fig. 19-10. It does not have a definite clock frequency in the
sense that **0** and **1** data pulses require the same time. A change is
made after each data bit and the width is used to distinguish between
0 and **1**. Decoding is done with a microprocessor (not a UART).

19-7 Pulse Width and Delta Modulation

A popular technique for transmitting analog signals is pulse width
modulation (Fig. 19-11). The carrier frequency and amplitude are
fixed, but the width of the pulse is proportional to the signal amplitude.
Here the carrier frequency is determined by the pulse generator. The
width of the pulse is controlled by a following monostable multivibra-

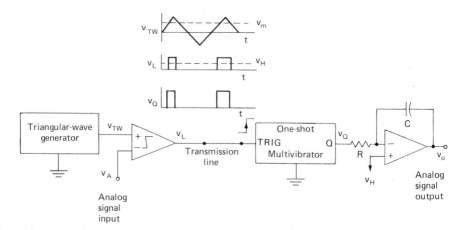

Figure 19-11 Pulse width modulation: (*a*) signal; (*b*) modulator; and (*c*) demodulator.

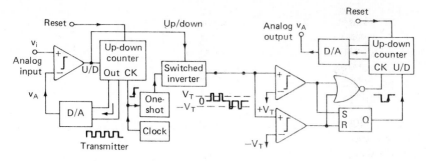

Figure 19-12 Block diagram of a delta modulation method.

tor. The pulse width of the timer is proportional to the control-voltage input, to which is applied the modulating voltage, assumed positive. The received pulse is restored to a standard level by a comparator, and the average of the pulse is extracted. A second-order Butterworth filter provides better removal of the carrier than a standard intergrator-averager (Chap. 10).

Delta modulation (Fig. 19-12), which might be classified as a combination of digital and analog techniques, is a method of transmitting changes in an analog signal. It is a variation of the servo method of A/D conversion (Chap. 13) in which a second up-down counter at the receiver end follows the transmitter counter. The transmitted signal consists of a positive pulse for an up-count and a negative pulse for a down-count. Because the maximum rate of change is limited to one least-significant bit per clock period, the response time (slew rate) is limited. It is best suited to analog signals which require high transmission accuracy but do not change rapidly with time.

19-8 Keyboard Encoders

A keypad or keyboard is a group of touch switches arranged electrically as a matrix. A switch encoder generates a binary code designating the switch number as its position in the array. One method of switch encoding is shown in Fig. 19-13 for 16 (or 12) switch array as used in telephone dialing or computer data entry. When a key is depressed, current will flow out of a specific Y line. The encoder produces the corresponding 4-bit number for each key. A signal provided when a key is pressed can be used to latch the encoder output. Switch debouncing is included.

Typewriter keyboard encoding is similar except for keyboard size. Usually the switch code is converted into ASCII (American Standard Code for Information Interchange). As Table 19-1 indicates, only 6 bits are needed to encode the standard 64-character (uppercase) keyboard.

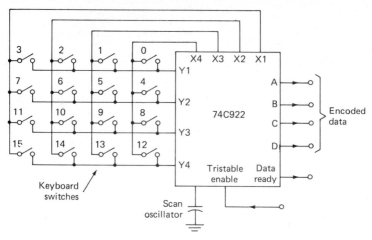

Figure 19-13 Keyboard encoder.

TABLE 19-1 ASCII Hexadecimal Equivalent*

			X_U					
	Control		Numeric		Uppercase		Lowercase	
X_L	0	1	2	3	4	5	6	7
0	NUL	DLE	Space	0	@	P	'	p
1	SOH	DC1	!	1	A	Q	a	q
2	STX	DC2	"	2	B	R	b	r
3	ETX	DC3	#	3	C	S	c	s
4	EOT	DC4	$	4	D	T	d	t
5	ENQ	NAK	%	5	E	U	e	u
6	ACK	SYN	&	6	F	V	f	v
7	BEL	ETB	'	7	G	W	g	w
8	BS	CAN	(	8	H	X	h	x
9	HT	EM	)	9	I	Y	i	y
A	LF	SUB	*	:	J	Z	j	z
B	VT	ESC	+	;	K	[	k	{
C	FF	FS	,	<	L	\	l	|
D	CR	GS	−	=	M	]	m	}
E	SO	RS	.	>	N	^	n	~
F	SI	US	/	?	O	—	o	DEL

*X_U is upper half byte, and X_L is lower half byte. Most-significant bit (D7) is 0 (parity ignored). Control characters are not printed. Example: code for T is $X_U X_L$ = 54 (= **01010100**).

If lowercase is added, 7 bits are required. The eighth bit may be a parity bit, but it is often ignored. Note that the BCD code for numerals is embedded in the code (add 3 for the higher half byte).

A number of control words are utilized for control, e.g., carriage return. Control words have zero in the upper 3 bits of the 7-bit code and can be formed by changing to zero the upper bits of another character. For example, control-M is CR (carriage return).

19-9   Dot Matrix Character Generator

Alphanumeric displays are often done by dot displays on an LED, LCD, printer, or CRT. All utilize an IC character generator to produce the desired characters. The process is most easily demonstrated by the LED alphanumeric display (Fig. 19-14). Basically the character generator is a read-only memory (ROM, see Chap. 20) in which the dot pattern of all characters to be generated is stored. One bit represents one displayed dot. To display all 64 uppercase ASCII characters on a 5 × 7 dot display requires a 2240-bit (minimum) ROM with a data word of 7 bits. The address is 9 bits, composed of the 6-bit ASCII-encoded input word and the 3-bit column address. A clock with counter and decoder scans the columns repeatedly. Only one column of the display is lit at any instant, but because of the rapid scan (minimum 30 Hz,

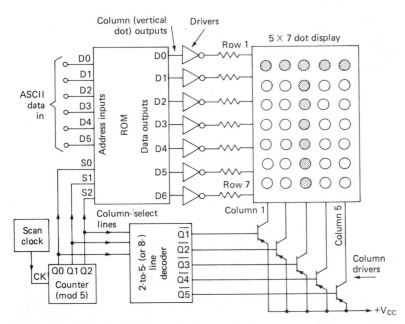

Figure 19-14 LED alphanumeric display with column-scanned character (single digit of a group shown).

usually much faster) the eye perceives the complete character. Driver and scan circuitry is similar to that for seven-segment display multiplexing (Chap. 17) except for the larger number of LEDs which must be scanned. If, as usual, multiple characters are to be scanned, the multiplexing is expanded to allow selection of displays, in effect extending the horizontal scan to $5N$, where N is the number of display characters.

Character generation for CRT displays employs basically the same method except that rows rather than columns are scanned. Because of the large number of characters and lines which must ordinarily be displayed, the circuitry is complex. With the standard refresh-type CRT terminals, the stored characters (data) are recalled from memory (RAM) during each high-speed scan of the screen while keeping track of the column and row number of each character. The output of the character generator during row scan is a rapidly changing serial digital signal which is also the video signal (beam intensity) for the CRT. Data sufficient for a full screen (typically 24 lines of 80 characters) are loaded into the RAM one byte at a time. Usually data (ASCII) are transmitted to the terminal in serial form and the RAM data input is a UART output.

Scanning of dot matrix LCD-type display requires trilevel drivers, but otherwise are similiar to CRT-type display drivers.

Dot matrix printers are column-scanned in the same manner. Impact-type printers employ seven thin solenoid-activated rods for printing in each column. A full line is printed across the page by controlling the impact timing (column position) after the motor reaches the desired speed.

Introduction to Microprocessors

A microprocessor is a simple but versatile computer in a single IC package. Its cost is low enough that even when accessory units such as memory are added, it is practical in electronic instruments of modest size. Because of the diversity of the microprocessors, the variety of ways of connecting the microprocessor to the instrument through various support ICs (hardware), and the programming variations (software), it is possible to discuss here only a small fraction of useful microprocessor functions and techniques. Furthermore, only two specific types of microprocessors, the Z-80/8085 and the 6805 will be described. The 6805 was chosen because of its simplicity, adaptability to instrument control, and low power requirements. The Z-80 and 8085 were chosen because of their popularity. All microprocessors have many features in common, and an understanding of one can be applied to others as well. In this sense, these are typical. Emphasis is placed here on dedicated microprocessor systems which are used in many instruments rather than in more general purpose systems such as employed in personal computers.

One purpose of this chapter is to provide an overall description of microprocessor operation for engineers and scientists who wish to communicate with engineers doing the detailed design of a microprocessor system. Another purpose is to provide the novice designer with a working knowledge of basic microprocessor hardware.

20-1 Microprocessor System Organization

A block diagram of a small microprocessor system is shown in Fig. 20-1. The heart of the microprocessor is the central processing unit

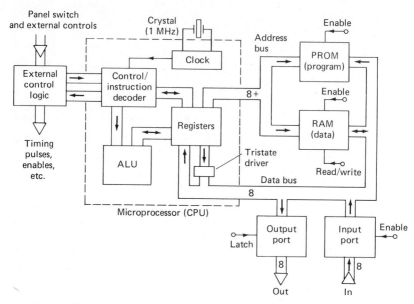

Figure 20-1   Simplified block diagram of an 8-bit microprocessor system.

(CPU), which consists of an arithmetic and logic unit (ALU), an internal control unit, and several registers. Arithmetic operations such as adding or comparing two numbers or bit manipulation (right or left shift) are done by the ALU. Included in the CPU group, loosely speaking, are the various timing circuits, drivers, and the clock. Sometimes the clock is external and can be divided into subcycles or phases. The data bus of this 8-bit microprocessor consists of eight separate lines which can transmit data either in or out. The 8-bit binary number represented is termed a byte. Bidirectional transfer of data requires a tristate output driver which can be put into the high-impedance state, in particular during an input phase. Data can be transferred to or from memory and input-output (I/O) ports. All these units are connected to the same data bus but remain inactive until turned on (enabled, selected, or strobed) by the microprocessor control. Additionally, in the case of memory units, only the part of the memory corresponding to the address code is activated. Normally the strobe pulse is brief, about one clock cycle, and the various units connected to the bus are enabled sequentially. At no time may two units be enabled to drive the bus simultaneously or the bus data will be completely invalid. This time-multiplexing arrangement allows many units to use the same bus without confusion.

Most operations initiated by a single program instruction require a number of clock cycles (typically three to eight) to complete. A number of cycles are required because each operation takes several steps, e.g., sending the address to memory, strobing the memory (so that the instruction is picked up by the CPU from the data bus), decoding the instruction, performing the indicated operation, and storing the result. Timing diagrams supplied by the manufacturer detail the process. Timing pulses also are sent by the microprocessor to the peripheral device to synchronize the operation.

Memory consists here of two parts, a PROM (programmable read-only memory), in which the program instructions are permanently stored, and a RAM (random-access memory), in which data are stored temporarily. Data stored in the RAM are lost when power is turned off, but data in a PROM remain. Nonprogrammable ROMs must be preprogrammed by the manufacturer by a masking process during the manufacture of the IC, an approach practical only for high-volume production. For the average user, electrically programmable ROMs or EPROMs are the only practical user-alterable ROMs. Alternatively, the RAM can replace the PROM in applications where (1) the program is loaded each time before each use (from a disk, perhaps), (2) the power is never turned off (battery standby), or (3) the program is frequently altered, as during initial trials. From the microprocessor's point of view, RAM and PROM are identical during read operations except for the address location which is specified (programmed and/or wired) by the user. Of course, only a RAM can respond to a write command. For the purposes of the following discussion, it will be assumed that the program instructions are stored in PROM, at a memory address which the microprocessor goes for the first instruction upon reset.

Memory allocation, i.e., the selection of the size of PROM and RAM together with address, depends on the specific task. The PROM size must obviously equal or exceed the number of instructions required.

Control logic external to the microprocessor of some type must be supplied if only to reset the microprocessor to its starting state. Halt, single-step, and direct-memory-access (DMA) modes might also be implemented. State codes and timing pulses are sometimes available to indicate the phase of an operation and allow synchronization of external hardware.

Actually the term *control lines* can be applied to all lines which are not data or address lines. Some control lines originate or terminate in ports which normally transmit data but are used instead to carry device-ready or other control information. Often microprocessor diagrams show a bus labeled *control*, but it must be recognized that it is a miscellaneous grouping rather than a homogeneous group, like the data and address buses.

Most operations required two distinct cycles, fetch and execute.

During the fetch cycle, a coded program instruction (often 1 byte) is retrieved from PROM and internally decoded (within the CPU). The address of the next instruction is kept track of by an internal program counter, which normally increments by 1 at the end of the fetch cycle. During the execute cycle, the operation set up by the previous fetch cycle is performed, e.g., the transfer of data from an input port to memory. Microprocessor operation consists mostly of a sequential series of regular fetch and execute cycles, occasionally broken up by operations with shorter or longer execute cycles or instructions which cause the program counter to jump to a different address.

20-2 Memory Organization

Memory for 8-bit microprocessors is divided into pages consisting of 256 bytes (Fig. 20-2). This is necessary because a 1-byte address (8 bits) can specify at most 256 address locations (0 to 255). Most microprocessors store a 2-byte address internally. Various microprocessors differ in the method of outputting the two address bytes to the memory. Most have an 8-bit (or 4-bit) address bus used for the high byte of the address and multiplex the lower byte of the address on the data bus (address first, then data). External hardware (latches) is required to restore the full address, as will be described subsequently. One way of looking at this organization is that the high address byte specifies the page

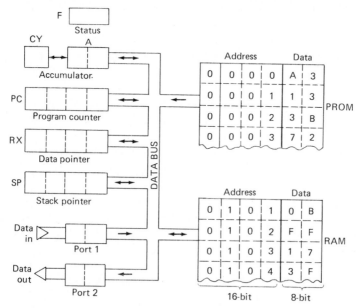

Figure 20-2 Simplified memory and register layout for a typical microprocessor.

number, while the lower byte specifies the line within the page. Recognizing the division into pages is important in understanding certain data-transfer instructions which can only transfer within a page because the part of the instruction specifying an address is only 1 byte in size. Some microprocessors (e.g., 1802) multiplex both the high and low bytes of the address on the 8-bit address bus (high byte first).

A microprocessor has an instruction set which can do addition, subtraction, transfer to and from memory, branching, and the like. Most sets number between 80 and 256 (for an 8-bit machine), but only 20 to 40 instructions are really distinct. Many instructions are minor variations of each other, such as leaving the number in the accumulator rather than in RAM following addition. Others incorporate a register reference number as a part of the instruction code so that a single type of instruction has 8 or 16 variations.

During the last part of the fetch cycle, the coded instruction is fetched from PROM and interpreted. After execution, the next instruction is fetched from the next higher address location sequentially. This continues until an instruction causing the program to stop is encountered or, more commonly, some type of branch to another part of the program or loop occurs. Usually the byte representing an instruction or datum as well as the addresses is represented by two hexadecimal numbers, although binary, octal, and decimal representations are occasionally used (see Table 20-1).

The 6805 and Z-80 have five main internal registers: an 8-bit accumulator which functions as a data scratchpad, a 12- to 16-bit program counter which points to the current instruction (PROM address), a data pointer or index register which points to data in memory, a stack pointer which points to the top address of a memory (RAM) storage area, and a status or flag register. In addition, there are a number of secondary registers which can be used interchangeably with the main registers.

TABLE 20-1 Decimal, Hexadecimal, Octal, and Binary Representation of Numbers 0 to 15

Hexa-decimal	Binary	Decimal	Octal	Hexa-decimal	Binary	Decimal	Octal
0	0000	0	0	8	1000	8	10
1	0001	1	1	9	1001	9	11
2	0010	2	2	A	1010	10	12
3	0011	3	3	B	1011	11	13
4	0100	4	4	C	1100	12	14
5	0101	5	5	D	1101	13	15
6	0110	6	6	E	1110	14	16
7	0111	7	7	F	1111	15	17

Only 4 bytes each of RAM and PROM are shown. These registers are 16- or 12-bits wide to allow addresses up to 16 bits (64 kilobytes = 65,536 bytes of data) to be referenced. Since the data bus is 8 bits, it requires at least two instruction cycles to move the 2-byte word in or out of the CPU. A carry bit (CY) necessary for various arithmetic operations such as addition is present as a part of the status register. It can be tested by various instructions or shifted to the accumulator but cannot be transferred into the data bus directly.

Within the CPU is a 13- to 16-bit counter program register PC, which keeps track of the location in PROM of the next instruction. Normally it starts with a specified location, for example, 0080, upon reset and is automatically incremented to the next address (or by two or three addresses for 2- or 3-byte instructions), unless a branch or jump instruction causes it to do otherwise.

The data pointer specifies a memory (usually RAM) address of data to be processed in the CPU. Subtraction of data at a specific address RAM location from that in the accumulator, for example, requires a pointer to that location.

The stack pointer specifies a memory address in a block of memory (RAM) or stack reserved for the temporary storage of data. Data are stored and later removed sequentially, starting from the top of RAM (highest address) in a first-in first-out order. Certain operations, such as the call to a subroutine and return, require the storage of addresses and other data in RAM.

Various status conditions are indicated in the flag or condition code register. Unlike the other registers which can be loaded or read directly by the program, bits in this register indicate the results of previous arithmetic or logic operations. One bit is the carry/borrow flag set by addition overflow or subtraction underflow. Another bit indicates when the accumulator is zero.

In the simplified architecture of Fig. 20-2, the considerable differences between the 8085/Z-80 and the 6805 are not shown.

20-3   Register Transfer Operations

Instructions are specified by a hexadecimal operation code, by name, and by a mnemonic (abbreviation or shorthand name). These op codes are at least 1 byte long and often are combined with one or two additional bytes to form the total instruction.

Transfer of data to and from registers within the CPU is an important software operation. Registers affected include the data pointer and program counter as well as general purpose registers. Consider first the Intel 8085 instruction set. It is a subset of the Zilog Z-80 instruction set, both of which are based on the earlier Intel 8080 set. Data may be transferred from the accumulator (A) to memory (M) and the reverse. Typical memory-register transfer instructions (8-bit load, BC register) are

Type	Mnemonic	Name	Operation	Op code
Intel	LD r, M	Load A from M	A ← M	0A
	LD M, r	Load M from A	M ← A	02

Transfer of data from on register r_s to another r_d is done by the instruction:

Intel	LD r_d, r_s	Load register	(r_d, r_s)	1$r_d r_s$

The Motorola 6805 instruction set is based on the older 6800 set. Transfer of data to memory from the accumulator using (direct addressing) uses the instructions:

Mot	LDA	Load A from M	A ← M	B6
	STA	Store A into M	M ← A	B7

Data transfer to the X register RX, which can function as a second accumulator, is similar.

Some instructions affect the individual bits within the accumulator data byte. An example is "shift left," which shifts bit D6 to D7, D5 to D6, etc. Bit D7 is transferred to the carry flag. Bit DO is 0 (if shift without carry), or the contents of the previous carry bit (if shift with carry). The left shift without carry instructions are

Intel	RLC	Shift left (rotate), with carry	D7→CY, CY→D0	07
Mot	ROL	Shift left, carry		49

20-4 Arithmetic, Logic, and Branching Operations

Arithmetic and logic operations are done 8 bits (plus carry) at a time, usually with the result in the accumulator.

To add a binary number in the memory location specified by the address stored in the data-pointer register RX (direct address) to the number in the accumulator A requires the following instruction

Intel	ADD M	Add memory to A	A ← A + M(RX)	86
Mot	ADD	Add memory to A	A ← A + M(RX)	BB

The result is stored in the accumulator. If a carry is generated, the carry flag CY is set. It is assumed that the desired 2-byte memory address has been placed RX by some previous instruction and that valid data are in the accumulator.

A related instruction is add immediate, which is 2 bytes long. It differs in that the 1-byte address of the number to be added to the accumulator is obtained from PROM, following the op code

| Intel | ADI data | Add immediate | A2 ← A + (byte 2) | C6 |
| Mot | ADD Imm | Add immediate | A ← A + M(RP) | AB |

The "immediate" means that the data follow the instruction in PROM. It can be considered a part of the instruction. A preliminary instruction which loads the data-pointer register is not needed. The notation "byte 2" means the second byte of the 2-byte instruction. Note that RP is the address or register number of the address where the data are stored. After retrieving the address, the RP is again incremented by 1, ready for the instruction.

Among other arithmetic instructions are subtract, add-with-carry (carry bit is used, as needed for upper byte of double-precision operations), and subtract immediate. These instructions are uncomplicated once the add and add immediate instructions are understood.

Logic operation instructions are similar in form to arithmetic instructions. An example is the exclusive-OR (XOR), which combines a number in RAM (address in RX) with the number in the accumulator. The instruction is

| Intel | XRA data | Exclusive-OR | A ← A V M(RX) | AE |
| Mot | ORA | Exclusive-OR | A ← A V M(RX) | BA |

Each bit of the two operands is computed, and the result is placed in the corresponding bit position in the accumulator. This instruction is useful in testing two data types for equality, because if they are equal, the result is all zeros in the accumulator. A subsequent instruction can be then used to detect the zero-accumulator condition. Other logic operations are exclusive-OR immediate and the standard and immediate forms of OR and AND operations.

An unconditional branch or jump to a new PROM address is done by instruction

| Intel | JMP | Unconditional long branch or jump | PC ← (byte 3) (byte 2) | C3 |

This instruction breaks up the normal retrieval of an instruction from the next PROM address indicated by the program counter RP. The next instruction is obtained from the address stored at address RP and

PR + 1 (2 bytes) and subsequent instructions from the new location of RP. In other words, the next 2 bytes in PROM following the op code are interpreted as an address rather than an instruction. The main value of this instruction is a program organization tool similar to the GO TO instruction of Fortran.

A conditional branch instruction is similar to the unconditional branch instruction except it takes place only when a specified condition of the status register occurs. If not, the next instruction is executed instead. When it is combined with arithmetic or logical operations, the result is similar to those performed by Fortran arithmetic IF statements. The conditional jump or branch instruction (3 bytes) which causes an address jump when the accumulator is zero is

| Intel | JZ | Long branch if A = 0 (Z = 1) | If Z = 1; then PC = (byte 3) (byte 2) | CA |
| Mot | BEQ | Long branch if A = 0 (Z = 1) | If Z = 1; then PC = (byte 3) (byte 2) | 27 |

Transfer of the program counter to the new 2-byte address, which is located immediately following the op code in PROM, occurs if the accumulator is zero; that is, if the Z bit of the status register is set. Otherwise, it goes to the instruction in PROM which is the next position beyond.

Another useful (short) branch instruction is quite similar to the instruction above except that transfer occurs if the carry bit (CY) of the accumulator is 1. This is indicated by setting bit C = 1 in the status register.

| Intel | JP2, NN | Jump if CY = 1 (C = 1) | If CY = 1, then PC = (byte 3) (byte 2) | DA |
| Mot | BC5 | | If CY = 1: PC = (Lo) ← byte 2 | 25 |

The Intel instruction is a long branch while the Motorola is a short branch.

20-5 Memory (RAM and PROM) Circuits

Of the numerous methods for storing digital data, the directly addressable semiconductor memory or RAM is almost invariably used for microprocessor applications. Other memories such as magnetic bubble or disk may supplement the RAM but do not completely replace it. A semiconductor RAM is a volatile memory, that is, data are lost when power is removed.

There is a division of memories into static and dynamic types. The

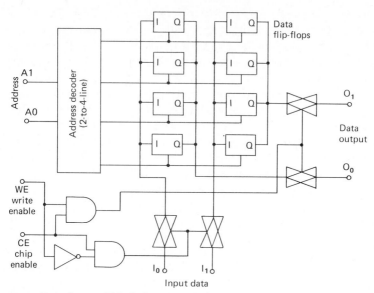

Figure 20-3 Internal block diagram of a very small 4 × 2) RAM.

lower-cost dynamic RAM requires a clock (frequency about 10 kHz to 1 MHz) to continuously recycle or refresh the data stored internally in small capacitors. If the refresh clock stops momentarily, the data are lost. Strobing of the output data ordinarily must be synchronized with the refresh clock so that it occurs during a nonmemory operation. Static RAMs have no clock or synchronization requirements. They are definitely easier to use and the discussion here will be restricted to them.

Memory size, or number of bits stored, is the RAM's most important characteristic. Invariably the number of bits is equal to some power of 2 in order to match the binary arithmetic of the microprocessor. A typical RAM is 1024 bytes, or 1 kilobyte in size, while a small RAM might have only 128 bytes. Bit or byte organization of RAMs differs. A RAM can be 1, 4, or 8 bits wide; i.e., each address would correspond to 1, 4, or 8 bits of data. For example, a 256 × 1 RAM would have 1 bit at 256 different addresses, while a 32 × 8 RAM would have an 8-bit byte at 32 different addresses (both are 256-bit RAMs). Most newer-design RAMs are 8 bits wide.

It is instructive to consider the internal block diagram of a RAM illustrated in Fig. 20-3. A special type of flip-flop holds each bit of data. When the local flip-flop enable line (from the decoder) is turned on, the flip-flop is connected to the data input and output lines through a transmission gate. Only when the input gate is turned on does the flip-flop follow the state (**0** or **1**) of the input. This occurs if the write-enable line (WE) is a logic 1 and the chip enable (CE) or strobe line is at logic 1. The flip-flop is thus latched. The two enabled flip-flops corresponding to the two input data bits (I_o, I_1) are latched simultaneously. Note that the outputs are in the high-impedance state during this write process. If the WE line is low, the outputs of the enabled flip-flops appear at the RAM output (O_o, O_1) when the chip is enabled (CE = 1). In this case the tristate drivers are not in the high-impedance state.

In microprocessor applications the input and output lines of the corresponding data bits are tied together and connected to the data-bus lines. The WE or read-write line from the microprocessor controls the direction of data flow along the bus. Note that in any case the logic shown will not allow simultaneous input and output data flow.

In memory-write operation, the address lines are turned on first because the address decoders exhibit appreciable delay. The time required for the address decoder to settle is referred to as the memory-access time, typically 50 to 500 ns. After the address has stabilized, the chip can be strobed by turning on the chip-enable (CE) line. During a write operation, if the CE line is turned on before the WE line, data from the RAM would appear on the data bus until WE is enabled. This would foul up the data bus because the CPU is also dumping data onto the bus. Thus it is good practice to switch the WE line before or simultaneously with the CE line. During a read operation the CE line can be turned on (with WE off) before the address lines settle, but the data will not be stable and should not be taken off the bus until after the access time delay.

Memory addressing for the 8085 is shown in Fig. 20-4. The low address byte is latched from the data bus at the proper time as determined by the address-latch-enable (ALE) line (falling edge). All 16 address bits are available during the last part of the instruction cycle when the data must be transferred to or from memory (see timing diagrams, next section).

Memory is enabled or selected through the chip select line CS. In the example shown, the address line A12 is used to switch between RAM and EPROM. When EPROM is on (address range 0000 to 0FFF, hex) the RAM is off. The 1K RAM has an address range of 1000 to 13FF, but since the decoding circuit is not a full decoder, the RAM will be also enabled from 1000 to 1FFF. The programmer must avoid the included duplicate addresses. A three-input NAND gate decoder disables the bank of memory for higher address (A13, A14, or A15 are 1) so that other devices may be connected.

Memory latching for the 6805 microprocessor is similar (Fig. 20-5) but differs in three main respects: (1) RAM (112 bytes) is internal; (2) only 13 bits of address (8K) are available; and (3) certain blocks of

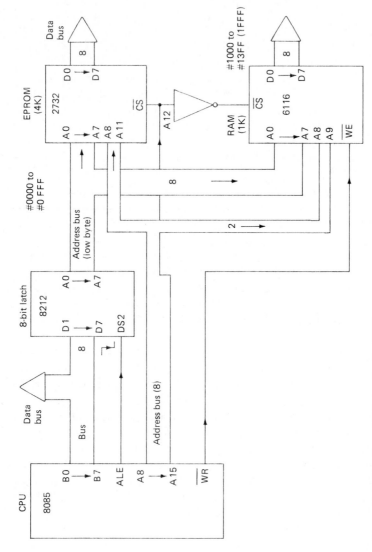

Figure 20-4 Memory addressing for the 8085.

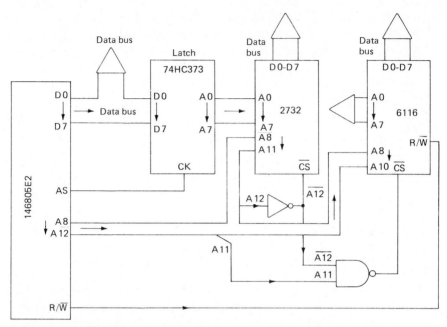

Figure 20-5 Connection of memory for the 6805 (MC146805E2) microprocessor.

memory are reserved for I/O and interrupt operations. The microprocessor and EPROM, as well as the support chips, are low-power CMOS
devices.

Memory mapping is somewhat complicated for the 6805 circuit. As
indicated in Table 20-2, the lowest block of memory is reserved for
memory-mapped ports (Sec. 20-7) and the next block for internal RAM.
The external decoding circuit disables EPROM for this address range.
The program memory (EPROM) starts at address 1000 hex (as determined by the reset vector). At the upper 16 bytes the memory range is
reserved for interrupt and reset vectors (Sec. 20-8).

It is suggested that a listing such as Table 20-2, which relates
software addresses to hardware addresses and memory allocation, be

TABLE 20-2 Memory Map Allocation Example

Address (hex)	Memory
0000–000F	PORTS
0010–007F	RAM (internal)
0080–07FF	Unused
0800–0FFF	RAM (external)
1000–1FEF	EPROM (program)
1FF0–1FFF	EPROM (vectors)

prepared at an early state in the microprocessor system development. Misunderstanding which could result, for example, in the programmer branching to a nonexisting area of memory (never-never land) is thus avoided.

The connection of a PROM and RAM to the microprocessor is almost identical except that only a RAM has a WE line. Some older-model PROMs require an additional negative power supply, an inconvenience.

Programming or writing into PROM is a comparatively difficult procedure. In all electrically programmable ROMs the bits are set or "burned in" by applying a higher-voltage pulse of short duration. This is done by a PROM programmer. Voltages and procedures are different for the various types and makes, and therefore a general purpose programmer must be rather complex, especially if high-speed and accuracy-checking circuits are included.

Erasable, electrically programmable read-only memories (EPROMs) are by far the most widely used PROMs (except in high-production-volume systems). They are composed of FETs with long-term, practically permanent charge storage on the gates. As with any FET, a voltage or charge on the gate can turn off the transistor. By trapping the charge on the gate in an electrically insulated region, the transistor can be turned off until the charge leaks off. With existing PROMs, it is estimated that the leakage time in the dark is many years: i.e., the transistor (1 bit) can be turned off practically permanently. When exposed to ultraviolet light, the insulator becomes slightly conducting, and under intense radiation the charge leaks off within a few minutes. All bits are thus returned to the discharge state (logic 1 usually). To charge the gate in the first instance requires the application of a voltage high enough (12 to 25 V) to cause an avalanche breakdown of the insulating region. By this process a bit is burned in. If a mistake is made, the entire PROM must be erased by exposure to ultraviolet light through a window on the top of the IC package. The programming pulses must be carefully timed to prevent damage to the PROM, and perhaps repeated if verification indicates that the bits have not been changed.

An electrically erasable programmable (Fig. 20-6) read-only memory (EEPROM) is a nonvolatile memory like an EPROM but, unlike an EPROM, data can be altered (erased and written) by the microprocessor. In this respect, it is like a RAM, but since the write time is long (10 ms/byte), it is too slow to replace RAM in most applications. The EEPROM is better thought of as an EPROM with an on-board programmer. The write command (WE) transfers data from the bus into an internal latch and starts the slow programming process. A busy (RDY/BSY) signal indicates when the programming is done.

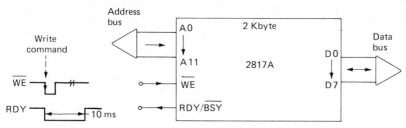

Figure 20-6 Electrically erasable programmable read-only memory (EEPROM).

20-6 Timing Diagrams

The sequencing of the various microprocessor events is best understood by timing diagrams. Generally they are rather complex because of the many peripheral devices (memory, ports, etc.) which must be controlled. The CPU generates the timing pulses from clock pulses to which all timing signals are referenced.

The number of clock cycles per operation for Z-80/8085 is not constant, as indicated in Fig. 20-7. Machine cycle M1 is an (instruction) fetch cycle which requires three clock cycles. M2 is also a memory-read cycle which here obtains the port code (low byte) from PROM, and during M3 the port code is transferred to the address lines while turning on the I/OM line. Other instructions have up to 5 cycles.

The timing diagram for the 6805 (Fig. 20-8) is rather similar but requires five clock cycles for the input instruction. Some other instructions require more clock cycles.

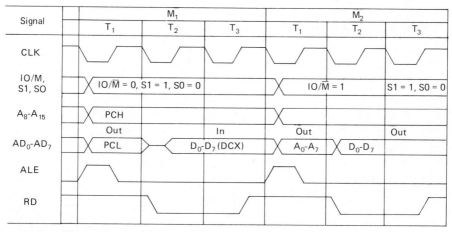

Figure 20-7 Timing diagram for the 8085.

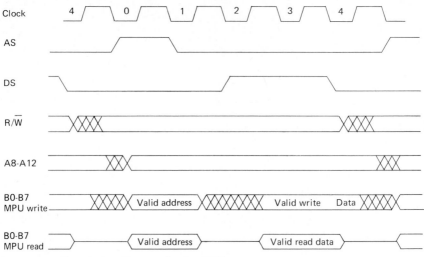

Figure 20-8   Bus timing diagram for the 6805.

20-7   Input-Output (I/O) Ports

The Intel 8080/8085 transfers data between the accumulator (or other register) and I/O ports in much the same way as it transfers data to and from memory. The technique of treating ports as a part of memory is referred to as a memory-mapped I/O. Two methods of separating RAM from ports are used; both involve turning on the appropriate enable line when the desired unit is to be activated. The first method (Fig. 20-9) is to turn on an I/O status bit (IO/$\overline{\text{M}}$). The status flip-flop output is brought out for control purposes, in particular to switch off the memory and turn on the port. A specific port is turned on if the I/O bit and the address line are on. Two instructions, which input or output data from port N to the accumulator are

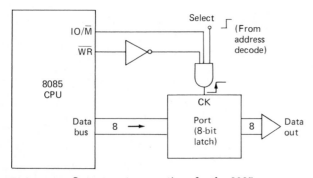

Figure 20-9   Output port connections for the 8085.

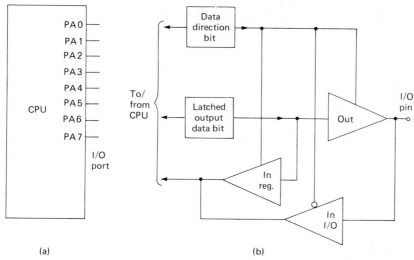

Figure 20-10 6805 port connections: (*a*) port lines on CPU; (*b*) internal logic for bidirectional line.

Intel	IN port	Input N	A ← (data)	DB
	OUT port	Output N	(data) ← A	D3

A second byte in the instruction is the port number N (8-bit).

An alternative method is to use the highest address bit A15 to switch between memory and I/O. It replaces the $\overline{IO/M}$ line. Presumably the required memory is small enough (under 32 kilobytes) so that there is no need for this bit in the RAM/PROM address.

The 6805 ports are also memory-mapped, but there is no I/O status line. Two 8-bit bidirectional ports are present on the version shown in Fig. 20-10. The direction (input or output) of each bit can be set separately under software control by a port direction register. The lowest memory addresses (0000 to 001F) are reserved for ports (others can be added externally).

Circuit details of two types of input ports are shown in Fig. 20-11. The simplest port is a set of eight transmission gates, which are turned on simultaneously by the strobe or enable pulse as supplied by the microprocessor at the proper time. This type of port is satisfactory if the data remain unchanged until after the port is read. The data might originate from a set of data switches. If the data change rapidly or the microprocessor may be delayed before reading the port, a latch type of port (Fig. 20-11*b*) is greatly superior. The latch pulse usually is supplied externally by the device generating the data, e.g., an A/D converter.

Eight type-D flip-flops with tristate outputs are available in a single

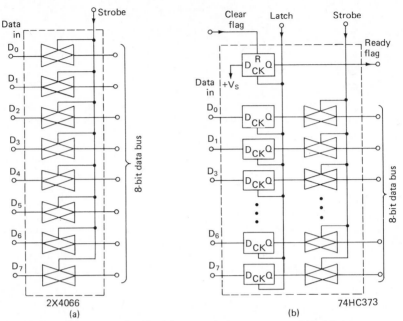

Figure 20-11   Input-port details: (*a*) transmission gate-type; (*b*) latch type.

IC package (Fig. 20-11*b*). A ninth flip-flop sometimes is added as a data-ready flag. It can be connected to a flag input or interrupt the microprocessor to signal the CPU that the port has data ready. After the input port has been read, the microprocessor must clear the ready flip-flop. One of the output ports can be utilized for this purpose.

A standard output port is shown in Fig. 20-12. It is essentially the same as the latched type-*D* flip-flop input port except that it lacks the tristate output (if needed the port of Fig. 20-11*b* can be used). A ninth flip-flop can be included to indicate to the external device that output data are ready (the device must supply a clear pulse). The output port

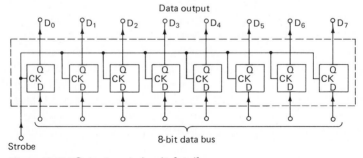

Figure 20-12   Output-port circuit detail.

may be constructed from a pair of quad latches. Ports as peripheral devices are discussed in Sec. 20–10.

20-8   Interrupt and Subroutine Transfers

The capacity of a microprocessor to exchange registers such as the program counter adds considerably to the programming flexibility. Calling a subroutine and the return to the main program with Intel software requires one instruction for each

Intel	CALL	Call	SP ← RP PC ← (byte 3) (byte 2)	CD
	RET	Return	RP ← SP	C9

Here SP is a stack pointer (a 2-byte register). The address of the first instruction of the subroutine RX is specified by bytes 3 and 2 of the instruction. Thus CALL puts the subroutine starting address into the program counter RP. Before doing so it saves the current contents on the program counter, which is later retrieved by the RETURN instruction, normally placed at the end of the subroutine program.

The 6805 also saves the current (calling) program counter 2-byte address on the stack and then branches to the starting address of the subroutine.

The subroutine call and return instructions are

Mot	BSR	Branch to Subroutine	M(SR) ← PC + (2 bytes)* PC ← PC + re/[†]	AD
	RTS	Return from Subroutine	PC ← M(SP)*	81

* Two bytes stored in stack: Sp is automatically decremental and incremental.
† re/ is relative address (2nd byte of instruction).

A 1-byte relative address allows transfer within 128 bytes of the calling address. Return is accomplished by putting the old address (after the call instruction) saved in the stack back into the program counter. Nested subroutines are possible.

Interrupt is a break in the main program initiated by an outside device to perform some high-priority operation. Interrupt is important, for example, in machine-control operations. It allows rapid response to sporatic data inputs, including alarm indicators. An interrupt process is initiated by a pulse applied to the interrupt (input) line of the microprocessor. After finishing its current instruction, a jump from the main program to an interrupt-service routine occurs. This is done by changing the program-counter register and is similar to a subroutine call.

One method of interrupt for the 8085 is to apply a signal to the RST

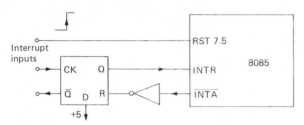

Figure 20-13 Interrupt connections for the 8085.

7.5 input (Fig. 20-13a). On the rising edge, an immediate branch to address 3C occurs, where a branch to the interrupt program is stored. It is equivalent to a hardware CALL instruction. Return to the main program is done by a RETURN instruction. Another interrupt input is INTR. A branch to an address placed on the bus by an external device occurs when it is high, at which time the output INTA goes low. Usually a flip-flop (Fig. 20-13) controls this handshake.

With the 6805, the 2-byte interrupt vector or starting address is stored in a specified location in EPROM (near the end of the memory range). Upon external interrupt, the old values of the program counter and the contents of the other CPU registers (total 5 bytes) are stored in the stack in a particular order. Then the interrupt vector is put into the program counter so that interrupt program is now executed. At the end of the interrupt program, the return instruction restores the contents of the registers to the values before interrupt. Regular program execution then continues. The interrupt and return instructions are

	(INT)	External Interrupt	M(SP) $\begin{cases} S \\ A \\ X \\ PC \end{cases}$		
Mot					
	RTI	Return from Interrupt	$\begin{matrix} S \\ A \\ X \\ PC \end{matrix}$	$\Bigg\}$ M(SP)	80

5 bytes stored in stack.

The 6805 also has a timer and software interrupt with a similar structure.

20-9 Programming Methods

The programming for a microprocessor is similar to programming in a higher-level language such as Fortran but is much more tedious because the small details of machine operation must be accounted for.

Initially an overall flowchart is made which lists in order the tasks to be performed. Major branches and loops are included, but details are ignored. Next the numbers and functions of registers and ports are assigned. To a large extent the choice is arbitrary, but of course once hardware is fixed (circuit wired), the software, such as port numbers, must correspond. Next the memory (PROM and RAM) addresses are assigned. For example, PROM might be allotted eight pages (2-kilobytes) of memory from 0000 to 07FF, while RAM might be allotted four pages from 0800 to 0CFF. It is assumed that the actual memory address lines are properly wired according to the chosen assignment, that is, the software and hardware coding must agree. When this is done, a detailed flowchart which references specific ports and memory locations by assigned number can be prepared.

Computer-aided programming is very desirable, if not necessary, for all but the smallest programs. If larger computers (IBM-PC, DEC, VAX, etc.) are available, one approach is to obtain a cross-compiler program intended for the specific microprocessor selected. The compiler accepts commands in mnemonic form, converts to op code, and allows branch locations to be specified as symbolic name. Absolute hexadecimal addresses are calculated by the compiler, which is a great convenience, especially when additions or corrections to the program are required. Various error checks are made, and the compiled program is listed (printed) in a convenient form with space for comments.

Often a simulation program is associated with the cross-compilation program. It carries out the program to be put in PROM in the same manner as the microprocessor but with a supervisory program to monitor the process. Should a problem develop, the steps in the calculation can be examined and traced easily. The simulator is intended to simulate the CPU and memory sections rather than a complete microprocessor system since it does not model the response of peripheral hardware. Successful simulation does not mean the microprocessor system will work, but if simulation is unsuccessful, the system certainly will not work.

A microprocessor system can be programmed to compile and simulate another microprocessor system, in particular the system under development. These development systems are desirable alternatives to cross-compilation and simulation by central computer.

Many smaller microprocessors do not have multiply or divide commands in the instruction set, but these operations can be done by combination of standard instructions such as addition and shift. Usually these operations are written as subroutines (perhaps 30 to 60 bytes long) to allow call from several parts of the program. These standard routines should be supplied by the manufacturer as a list of instructions or perhaps as a preprogrammed ROM. Other standard routines include binary-to-BCD conversion (and the reverse) and double-precision addition and subtraction.

20-10   Peripheral Devices

Almost any integrated circuit in a microprocessor system other than the CPU can be considered a support or peripheral device. Usually main memory is not considered peripheral while expansion memory may be. Only a few selected peripherals will be discussed here since the interfacing of peripherals does not differ greatly from one device to another, at least within a microprocessor family.

One useful support device is the (programmable) peripheral interface adapter (PIA), a type of general purpose port (Fig. 20-14). They usually have two or three 8-bit ports, at least one of which is bidirectional. Various programmable modes may include conversion from input to output lines and the use of some lines for strobes or handshaking. Interrupt request lines are often available for the input ports. The existence of programmable options means that various software control words must be loaded into specific memory locations before use. Versatility is thus achieved at the expense of convenience. Often RAM and a pair of 8-bit memory-mapped ports are combined into a single chip. When tailored to a specific microprocessor, the connection to the CPU is simple.

External programmable counter/timers are useful if, unlike the 6805 model described, there is no internal timer. Most counter peripherals have two to four counters which can be controlled and read by the CPU. Besides pulse counting, a standard option is counting the CPU

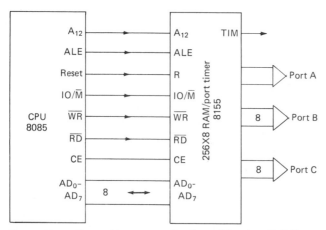

Figure 20-14 Connection to an Intel combination RAM/port peripheral.

clock pulses (divided by 2^N, N programmable) for an interval determined by an external signal. Thus either period or frequency counter functions can be implemented. Some devices include RAM or I/O ports also.

Other common peripherals are A/D, and D/A converters (Chap. 13), UARTs or other serial communication interfaces, display drivers, keyboard scanners (Chap. 19), CRT controllers, direct-memory-access (DMA) controllers, and floppy disk controllers.

20-11   Microprocessor Families

Microprocessors can be classified by data bus size (4, 8, 16, or 32 bits), fabrication technology (thus speed and power consumption) and family (primary developer/manufacturer). Popular microprocessors fall into four to six families, as indicated in Table 20-3. The Zilog Z-80 uses an improved version of the Intel programming language originally introduced by the 8080. The NSC800 as well as more advanced 16-bit machines also use this language. Another widely used language was introduced by Motorola with the 6800 and adapted by several manufacturers. The 6502 used in Apple computers, for example, has a very similar, but not identical language. RCA developed a substantially different architecture and language for the 1802 series. As the first low-power CMOS microprocessor, it was and is used in many instruments. The language is less powerful than Intel or Motorola but RCA has introduced higher-level language packages to overcome this problem.

Most microprocessors and memory currently are fabricated using NMOS. Power-supply current drain is substantial (50 to 100 mA per chip) but speed is fairly high (2 to 8 MHz clock frequency). A few bipolar microprocessors are available which have higher speed (20 MHz) but require still higher power. There is a trend toward CMOS-type microprocessors (such as the 146805) which require much less supply current (1 to 5 mA) and thus are suitable for battery operation. Many CMOS versions of the original NMOS microprocessor (i.e., Z-80C) are available.

Sixteen-bit microprocessors have substantially higher speed and

TABLE 20-3   Microprocessor Families

Family	Original
8085	Intel 8080
Z-80	Intel 8080
6805	Mot 6800
6502	Mot 6800
1802	RCA 1802

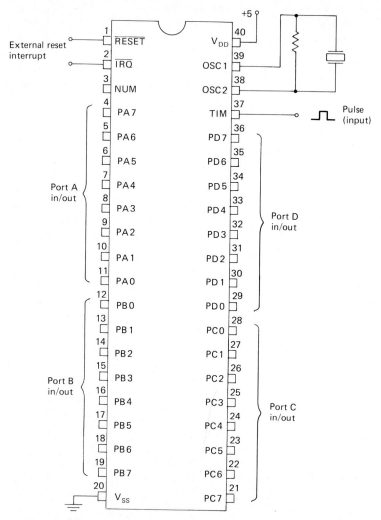

Figure 20-15 Microcomputer version of the 6805 (MC146805G2).

computing power than 8-bit machines. Equal accuracy is possible with an 8-bit machine using double- (or higher-) precision routines, but many more instructions and thus much more time are required. Most 8-bit machines multiply and divide via lengthy software subroutines, while 16-bit machines have fast hardware routines. Larger memory size and better data transfer instructions are other advantages of the larger machines.

However, 8-bit microprocessors are in most cases best suited for dedicated instrument control because the higher speed and precision of the larger machines may not be needed. The lower cost of the microprocessor components and less wiring needed for 8-bit machines are substantial advantages for smaller systems.

Four-bit microprocessors are used in low-cost high-production-volume applications. The greater difficulty in programming and the high cost of the development systems make the 4-bit microprocessors undesirable for low-volume applications.

20-12   Microcomputers

Microprocessors which have all the necessary components (CPU, RAM, ports, and ROM or EPROM) within the same chip are termed *micro-computers*. These single-chip devices are the major part of many industrial and consumer electronic controls.

The single-chip version of the 6805 is shown in Fig. 20-15. Four 8-bit bidirectional ports and a counter/timer are the only way of transferring data in and out. No data or address bus lines are brought out. For high-volume applications, a ROM is prepared by the manufacturer from data (op-code list) supplied by the customer. The program in ROM is permanent and therefore great care must be taken to produce an error-free program. ROMs are cost-effective only in high production volumes (over 1000 to 5000 units).

Microcomputers with internal EPROMs are available (at a price) for small production quantities and for system development. Programming of the EPROM may be difficult if a specialized programmer is required.

DATA SWITCHING, CONTROL, AND READOUT DESIGN EXAMPLE

Example 6   Digitally Controlled Timer

DESCRIPTION: A time pulse selected by a three-digit thumbwheel switch is produced when a switch is pressed.

SPECIFICATIONS
  Resolution: three digits (± 0.01 s)
  Interval: 0 to 9.99 s*
  Output: TTL-compatible (0/3.6 V)
  Trigger: Push-button switch (or TTL pulse)

DESIGN CONSIDERATIONS: The design is based on a three-stage decade down counter connected to a clock generator (555). A clock produces a stream of 10-ms pulses. The counter is preset to N (12-bit BCD number) by a brief pulse (PE) when the start switch is pressed. The **0** output of the 4522 is high when the counter (all four Q outputs) is zero provided the CF input connected to the **0** of a previous state is high. Upon preset, **0** therefore becomes low and T_{out} becomes high. During the

———————————

* Proportional to clock frequency, which is easily changed.

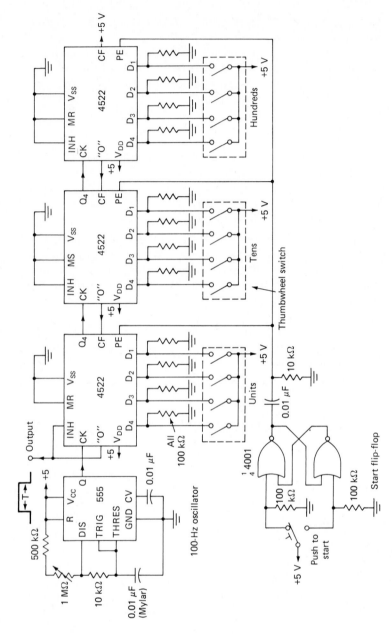

Figure E-6 Digitally controlled interval timer.

count process the 0 output of the most-significant digit goes high first, thus allowing, through the cascade connection (CF), the next-most-significant 0 to go high when it reaches a zero count (otherwise it resets to 9, as usual for a decade counter). After the least-significant digit reaches zero after all other digits are zero, its 0 goes high and T_{out} returns to zero, ending the timed interval.

Other interval times can be obtained by changing the clock frequency. Interval accuracy depends on the frequency stability. For the 555 timer, it is the 0.1 to 1 percent range. A 60-Hz timer with a six-digit counter might be preferred for longer times (0.1- or 1-s resolution).

DATA SWITCHING, CONTROL, AND READOUT DESIGN PROBLEMS

25 Design a push-button-switch-activated timer which will turn on a 115-V light (100 W) for 3 s. A 12-V relay (coil resistance of 300 Ω) with 2-A contacts is available. One contact of the switch is grounded.

26 Design a circuit which will deliver an output pulse approximately 1 μs wide after counting 100 input pulses. Assume that TTL-compatible logic is used and that a clear pulse occurs before the input pulses start. Further input pulses must not result in another output pulse (unless a clear pulse occurs).

27 Give a circuit a PLL type of frequency multiplier with a multiplication factor of 200. Assume a square-wave input F_i in the range 0.3 to 1.5 Hz. Use a 565, 4046, or similar IC.

28 Design a three-channel analog time-multiplex system which requires only one line (one wire plus ground). The input signal is assumed limited to ±5 V full-scale. A fourth channel may be utilized for synchronization, e. g., higher than full-scale voltage as detected by a comparator.

29 A three-digit frequency counter using the 60-Hz line as a time base is required. The maximum frequency is 9.99 kHz, and a new reading (display update) must occur every 0.2 s or faster. A latched output is suggested.

30 Design a digitally controlled interval timer based on a presettable down (or up) counter. It is triggered by a brief (< 0.1-ms) start pulse. Assume a clock input frequency of 1 kHz and a maximum time of 99 ms (resolution 1 ms). The time should be set by a two-digit thumbwheel switch.

31 Give a complete circuit for a pulse counter with latch and multiplexed four-digit display.

32 Design a tone decoder that will trip when frequencies of 700 and 1000 Hz are present.

33 Devise the circuits to transmit and receive a digital signal over a single narrow-band line or link. The frequency response is 100 to 3000 Hz, and the attenuation varies between -3 and -10 dB. Data rates will not exceed 10 b/s.

34 A tape-recorder-controlled slide projector is required. When the slide is to be advanced, a 1-kHz tone 1 s long appears in the recording, which otherwise consists of speech (describing the slides). To advance a slide, a relay contact must be closed for 0.2 s minimum. Devise a circuit which can be attached to the tape-recorder speaker output (1 V peak to peak minimum for tone) to accomplish this.

35 A remote light switch (relay or triac) controlled by a 30-kHz signal transmitted on the power line is required. Assume that the signal will be attenuated by a factor of 10 and that at this frequency the line shunt impedance is over 30 Ω and interference (noise) under 100 mV peak-to-peak. Design a system to accomplish this task.

36 Give a circuit for a port that will read two BCD digits from a thumbwheel. Next expand the circuit so that it will read in six digits.

Power Circuits

21

Power Supplies

Power supplies are often taken for granted, and with good reason: most supplies generally deliver a pure, regulated dc voltage as expected and without trouble. For many or most instruments the convenience and modest cost of a commercial modular power supply make it the best choice. Those who make this choice will consult this chapter only to clarify the terminology involved in power-supply specification. However, power supplies are not difficult to construct and often cost less when put together by the user, especially if multiple output voltages or other nonstandard features are needed.

21-1 Power-Supply Characteristics

A single power supply is a constant dc voltage source and is equivalent to a battery with an output voltage of V_s. Either the positive or negative terminal may be grounded, but for most applications discussed in this text, especially logic circuits, the ground is negative, as indicated in Fig. 21-1a. A dual supply is the equivalent of two batteries in series with a common connection (Fig. 21-1b). Almost invariably the common connection is grounded when used to power op-amps.

The output voltage V_s of a supply is its most important parameter. Fixed-voltage supplies, that is, ± 15 or ± 5 V, are most convenient and least costly for instrument power, although variable-voltage supplies, for example, 0 to 20 V, are often convenient for test purposes.

Maximum output current is the next most important parameter. Presumably the set output voltage will change insignificantly under load, i.e., as current is drawn, until the current limit is exceeded. Most all regulated supplies now are protected by a circuit which limits the current to some safe value, perhaps 10 to 50 percent above the stated maximum. The current limit is effective for practically any regulated

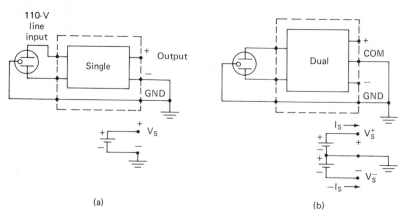

supply when short-circuited for brief intervals, but not all supplies can withstand a continuous short without overheating and damage. Better supplies have a thermal shutdown in addition to overcurrent protection. For long life a supply should not be operated continuously more than 70 to 90 percent of its current capacity.

Although most op-amp and logic circuits are rather insensitive to power-supply fluctuations, good voltage regulation is always desirable and often essential. Parameters characterizing regulator performance are regulation against load, regulation against line, ripple, and precison. Regulation against load refers to the change (reduction) in voltage ΔV_s as current I_s is drawn. It is often specified as the fractional change in output voltage when loaded. Another way of specifying this is the effective internal (Thevenin) resistance of the supply. Regulation against line is the percentage change in output voltage for a given percentage change in line voltage.

Ripple and/or noise refers to a presumably small ac component which is generally present in addition to the desired dc component. Ripple has frequency components which are multiples of line frequency (mostly 60 and 120 Hz). A good but not exceptional regulator will have a ripple content no higher than 20 mV rms when measured at full load.

Precision refers to absolute accuracy or difference between the stated voltage and the actual voltage. For most applications, in fact, it is important that the supply voltage remain constant, but the exact value is not critical. Precision supplies (see Sec. 11-8) are needed; however, where a reference voltage is assumed, as in some digital voltmeters and A/D converters.

21-2   Rectifier-Capacitor Input Section

Nearly all line-operated instrument power supplies use a stepdown transformer to provide the proper voltage and the necessary isolation between the instrument ground and the power line. Current passes through the diode (rectifier) to charge the capacitor (Fig. 21-2). The peak voltage to which the capacitor is charged is equal to the peak voltage of the transformer secondary less the voltage drop across the rectifier (~ 0.7 V). Since transformers are rated in rms volts at full load, a standard 12-V transformer will deliver 16.8 V peak—more if lightly loaded. Transformer current ratings are also stated in rms but apply to a resistive load and must be derated (especially low-cost units) because the current is drawn from the transformer in pulses. Power loss is proportional to the current squared, and therefore the transformer heating will be substantially higher for a given average current if the current is pulsating. In practice, the maximum direct current available is about 40 to 80 percent of the transformer rating.

It should be pointed out that power-supply parameters such as ripple are difficult to calculate accurately because the exact characteristics of the individual components, e.g., transformers, are usually not known. Thus calculations of the precise magnitude and waveshape of ripple under particular conditions are of little value. All that is really wanted are estimates or limits of supply ripple or regulation. Instead of going into the details of power-supply analysis, reliance will be made here on practical rules and graphs.

Full-wave rectification (Fig. 21-3) is desirable except for small load currents. The two main advantages are that the input capacitor C_I need only be half as large because the discharge time is half as long (frequency twice as high), and the transformer loss is less when current is drawn for both the positive and negative cycles. Transformer

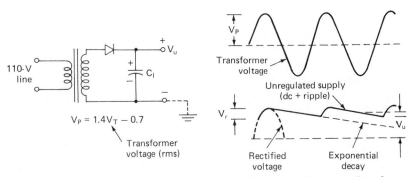

Figure 21-2 Power-supply input sections: (*a*) half-wave rectification; (*b*) voltage input and output; (*c*) ripple analysis.

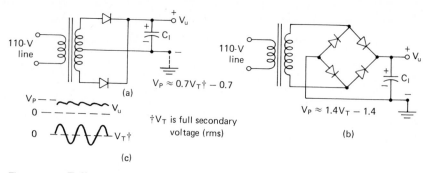

Figure 21-3 Full-wave rectifiers: (*a*) center tap; (*b*) bridge rectifier; (*c*) waveform.

loss is higher for half-wave rectification not only because the peak current is higher but also because the net direct current which flows through the secondary can cause magnetic saturation of the transformer core. These losses are not very important if the transformer is lightly loaded. Another advantage of full-wave rectification is that the high-voltage surge which often occurs when the transformer is switched off is harmlessly discharged through the rectifiers to the capacitors (for half-wave rectifiers the surge across the diode may exceed the inverse breakdown voltage and destroy the diode).

The two methods of obtaining full-wave rectification are shown in Fig. 21-3. Where a suitable transformer is available, the center-tapped circuit is preferred because it can be used in dual supplies (see Figs. 21-6 and E-7).

Current I_L is drawn from the supply by the load, which is represented here by an equivalent resistor R_L. It is the discharge of the capacitor C_I through R_L which causes the voltage drop between cycles seen on an oscilloscope as a sawtooth or ripple. Normally, the power-supply input-section voltages are calculated when maximum current is drawn, since this corresponds to the worst case (maximum ripple). To a fair approximation, the maximum direct current drawn from the complete supply I_L, even with regulation, is equal to that drawn from the input section, and therefore the effective load resistance R_L is equal to V_u/I_L where V_u is the unregulated voltage (dc or average value). Between cycles (half cycles for full-wave) the capacitor discharges exponentially with a time constant of R_LC_I (see Fig. 21-2). Between cycles the unregulated voltage V_u is roughly

$$V_u = V_\mu e^{-t/RLC_s} \approx V_p \left(1 - \frac{t}{R_LC_I} \right) \tag{21-1}$$

where the binomial approximation has been used to obtain the second expression. It is valid for a small voltage drop between cycles, i.e., for large values of C_I. In terms of this equation, the average or dc unregulated voltage is

$$V_u \approx V_P \left(1 - \frac{T_c}{2R_LC_I} \right) \tag{21-2}$$

where T_c is the period between cycles ($T_c = \frac{1}{60}$ for half-wave and $T_c = \frac{1}{120}$ for full-wave).

The peak-to-peak ripple v_r from this equation is

$$v_r \approx \frac{V_pT_c}{2R_LC_I} \tag{21-3}$$

The input capacitor C_I must be large to reduce the ripple to an acceptable level but should not be larger than necessary because these capacitors are costly and bulky. An acceptable compromise is to choose a capacitor such that the ripple (peak to peak) is between 5 and 20 percent under full load. Selection of a capacitor is simplified by the graph and formulas of Fig. 21-4. To use this graph, the maximum load current I_L and peak voltage V_P must be estimated. Note that usually V_P is only slightly higher (2 to 10 percent) than the unregulated voltage V_u and only slightly lower (5 to 15 percent) than the peak transformer voltage. For example a 12.6-V rms transformer into a full-wave bridge rectifier will produce a dc voltage V_u of 16 V (10 percent ripple). A load current of 50 mA would require a 200-μF capacitor for a 10 percent ripple.

Occasionally a voltage-doubler rectifier circuit (Fig. 21-5) is used if a

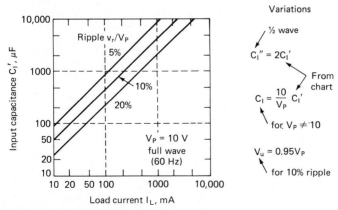

Figure 21-4 Input-capacitor selection graph. Note scale factor for peak voltages V_p other than 10 V.

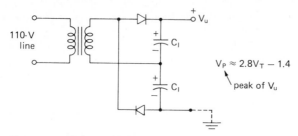

Figure 21-5 Voltage-doubler circuit.

transformer of sufficiently high secondary voltage is not available. The disadvantages of half-wave rectification are magnified, and the circuit is suited primarily for light current loads. The capacitors C_I are four times the value calculated by the graph of Fig. 21-4.

Inductor input-filter supplies are unpopular because of the additional cost and weight of the inductor. They are more efficient, an advantage for high-power applications, but otherwise have little to recommend them. However, inductors are used in switching regulators (Sec. 21-10). In this case the inductors are small because of the high switching frequency.

21-3 Unregulated Supplies

An unregulated supply can consist of just the transformer and input section. Sometimes a ripple filter is added. In Fig. 21-6 a complete unregulated dual (0-V) supply is given. It is designed by the methods discussed above. Such a supply is suitable where unregulated power is required (lamps, relays) or as an input to a voltage regulator.

Sometimes a passive RC filter section is added to reduce ripple. It should be pointed out, however, that ripple reduction is accomplished more effectively with IC voltage regulators, and RC filters are not routinely useful. An appropriate filter is shown in Fig. 21-7. It is simply a low-pass filter intended to reduce the dc voltage only slightly (V_u to V_u') while reducing the ripple greatly (v_r to v_r'). The reduction in

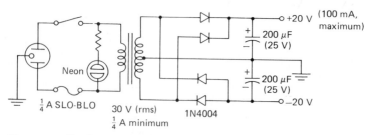

Figure 21-6 Dual unregulated power supply.

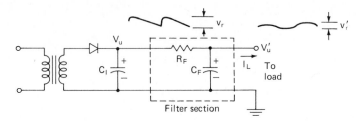

Figure 21-7 *RC* power-supply filters.

dc voltage $V_u - V'_u$ due to the load current I_L is calculated from Ohm's law. Note that the voltage is load-dependent, i.e., the load regulation is poor. It is difficult to obtain an attenuation factor of more than 5 or 10 at even modest load currents without large capacitors or poor regulation.

21-4 Zener Diode Voltage Regulators

Zener or voltage-regulator diodes are the basis of nearly all regulated supplies. The simplest regulator circuit consists of a Zener diode and limiting resistor (Fig. 21-8). As stated in Chap. 2, a Zener diode is normally reverse-biased into the breakdown region, which occurs at a particular voltage V_Z. The equivalent circuit of the Zener diode biased in this region is a battery V_Z with a small series resistance R_Z. Current from the unregulated supply $I_u = (V_u - V_Z)/R_d$ which flows through R_D divides between the Zener I_Z and external load I_L. As the load current changes, the Zener takes up the difference. Load current must be limited or the diode current will drop to zero and stop regulating.

The Zener-current change for a given line-voltage change is rather large. Care must be taken in the circuit design to ensure that the Zener current never becomes excessive or drops to zero for any normal combination of line-voltage changes, load-current changes, and component tolerance. Note that these variations are magnified if the

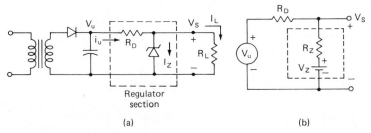

(a)   (b)

Figure 21-8 Zener diode voltage: (*a*) power-supply circuit; (*b*) equivalent circuit.

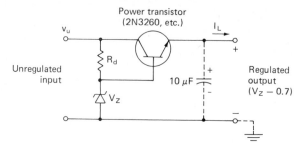

Figure 21-9 Emitter-follower Zener voltage regulator. Current overload not provided.

difference between the unregulated voltage and regulated voltage is small.

Zener diodes have maximum power dissipations $I_Z V_Z$ ranging from 100 mW to over 3 W. Their life expectancy is enhanced considerably by operating below half rated power. Higher-power diode regulators are not used since regulators with transistors or ICs cost less and perform better.

A variation on the diode regulator is the Zener-diode-transistor regulator shown in Fig. 21-9. The voltage across the Zener is calculated as above. It is applied to the base of the transistor connected as an emitter follower, so that the output (emitter voltage) is equal to the Zener voltage less the base-emitter drop (0.7 V). Because the base current is relatively small (I_L/β), the Zener current does not change as much as the circuit of Fig. 21-8. A further advantage is that the power is dissipated in the power transistor, a less costly device than a power Zener diode. A disadvantage of this circuit is a lack of short-circuit-current limit and medicocre regulation.

21-5   Op-Amp Regulators

Voltage regulators based on op-amps are described here as a basis of understanding the series-regulator circuit operation in general. One such regulator is shown in Fig. 21-10. A Zener-diode regulator provides the voltage reference V_Z. Any voltage difference between this and the potentiometer output V', a fraction α of the output voltage V_s, is amplified by the op-amp. Its output V_c drives the base of Q1. Normally Q2 is cut off and can be ignored. The series transistor must have high power and current characteristics. Often Q1 is replaced by a Darlington transistor for higher current gain. Since Q1 is effectively an emitter-follower (and the drop across R_1 is small), the regulated output voltage V_s will equal the base voltage less the usual base-emitter diode drop (0.7 V). Negative feedback adjusts V_c, and thus V_s,

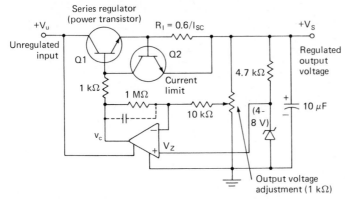

Figure 21-10 Series voltage regulator, op-amp version.

so that the op-amp input voltage difference $(V_Z - \alpha V_s)$ is small; that is, $V_s - V_Z/\alpha \approx 0$. This is the basis of the voltage-regulator action. Note that α is less than unity for this circuit, which means that the Zener voltage must be less than the output voltage.

As with other active voltage regulators, good ac as well as dc regulation is provided. The reduction in ripple is much more effective than with passive RC filters. Care must be taken to ensure an adequate voltage difference (0.6 to 1 V) across Q1 at all times, particularly during the ripple-cycle minimum, or the transistor will saturate. Inadequate regulator input voltage is seen as a negative notch in the output, or dropout of regulation, synchronized with the line frequency when the regulator is heavily loaded. Should this occur, the transformer voltage or size of the input capacitor must be increased.

The output capacitor (electrolytic or tantalum) provides a low impedance at higher frequencies where the regulator effectiveness decreases. Logic and high-frequency (megahertz) circuits require another parallel ceramic or paper capacitor (0.01 to 0.1 μF) near the circuit (not at the power supply) for adequate bypassing.

Short-circuit or current-overload protection is provided by Q2. Normally the voltage drop across the low-value resistor R_I in series with the load is so small that Q2 is off. Only when a high load current flows will the voltage drop exceed the emitter-base diode voltage (0.6 V). In this case the collector current will flow through Q2 and Q1, causing the output voltage to be reduced. Under short-circuit conditions $(I_L = I_{SC})$, the op-amp loses control, the voltage drop across R_I becomes constant at 0.6 to 0.7 V, and the current is limited. The current limit protects Q1 and perhaps the external circuit, and adequate cooling must be provided to allow for this possibility.

A low-current low-voltage regulator especially suited for battery

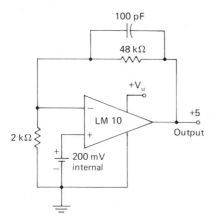

Figure 21-11 A low-voltage, low-power regulator.

operation is shown in Fig. 21-11. It takes advantage of the internal regulated voltage (200 mV) of the LM10 which is simply amplified to the desired level by the op-amp in the noninverting configuration. Voltage drop is small since the output can go nearly to the positive supply, but the current limit is low (5 to 10 mA).

21-6 Overvoltage Protection

Sometimes overvoltage protection is desirable. Application for a short time of a voltage a bit (10 percent) above the maximum ratings is fairly safe, but a factor-of-2 overvoltage is likely to destroy the semiconductor device. All the ICs in an instrument can be wiped out by a transient voltage surge, a worrisome possibility, especially in large and expensive instruments. To understand how this might happen, it must first be realized that the usual series regulator responds to overvoltage by turning off the series transistor. Since series regulators do not act as current sinks, they passively allow the output voltage to rise unchecked if current is injected from an external source. In instruments which have several supplies, an accidental cross connection may cause the output of a lower-voltage supply to rise to that of the higher-voltage supply. Another way of introducing an overvoltage (perhaps 110 V) is through an external input connection. The input IC will immediately blow and may short the input through the destroyed IC to the power supply. Without overvoltage protection, all the remaining ICs connected to the supply and the supply itself may be destroyed.

A simple and effective overvoltage protection circuit is the SCR "crowbar" circuit of Fig. 21-12. Normally the SCR is off and has no effect on the regulated supply. Should a voltage surge exceed the Zener

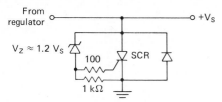

Figure 21-12 Overvoltage protection circuit.

voltage, current will flow through the Zener into the SCR gate and turn the SCR on, thus shorting the power supply. The SCR remains on until the power supply is switched off. The SCR current rating should exceed the maximum surge current expected. Negative surges are limited by the reversed diode. Negative supplies can be protected by the same circuit by reversing the leads.

21-7 Voltage-Regulator ICs

Voltage-regulator ICs are effective, convenient, inexpensive, and therefore popular in electronic instrument power supplies. Basically they are equivalent to the op-amp voltage regulators described above except that the circuit is largely incorporated into one package. Only a few of the many units available will be discussed here.

The simplest units are the three-terminal fixed-voltage regulators (Fig. 21-13). Popular sizes are 5, 12, and 15 V. All that is needed is to connect the input and common to an unregulated supply, and the regulated voltage is taken from the output. Most regulators require that the input voltage be at least 2 or 3 V higher than the output. Current limit (short-circuit protection) is provided. When operated without heat sinks, as is usually the case, the power dissipation $I_L(V_u - V_s)$ is limited to about 1 W typically for cases with heat-sink tabs and 100 mW for the low-power (L) versions. These units are therefore most satisfactory for low-current applications. The unit shown is a positive

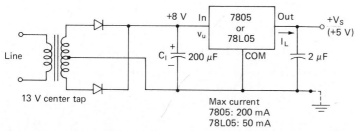

Figure 21-13 Power supply utilizing a three-terminal IC positive voltage regulator: (*a*) fixed; (*b*) variable.

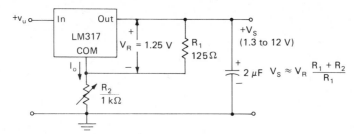

Figure 21-14 Adjustable regulated supply.

regulator (input and output voltages are positive with respect to the common terminal). A series of negative three-terminal voltage regulators is also available (examples, 7905 or 79L12).

Adjustable regulators such as the 317 (or 317L, Fig. 21-14) can be set to any voltage between 1.25 and 35 V. A resistor pair (R1, R2) divides the regulated output voltage (V_s) down to the reference level (1.25 V) which appears across the output and adjustment (common) terminals. The somewhat variable current flowing out of the adjustment terminal I_A must be taken into account in the voltage ratio calculation.

Actual average current ratings for regulators are determined primarily by the allowable power dissipation, which in turn depends on the regulator voltage drop and heat sink size. It is a mistake to assume a 1-A regulator will deliver anything close to 1 A continuously. From the power dissipation viewpoint, it is usually not desirable to combine a low-voltage (5-V) higher-current regulator with a higher-voltage (15-V) regulator by supplying both with the same unregulated voltage.

An older but versatile regulator is the 723 chip. It is especially useful as a regulator with an external power (series pass) transistor, as indicated in Fig. 21-15. Separate power transistors are relatively

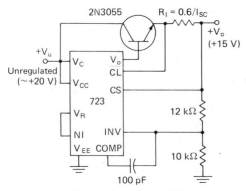

Figure 21-15 External-transistor IC voltage regulators (positive output).

inexpensive and easy to cool with a heat sink. External transistors are practically indispensible for higher current supplies. The main disadvantage is the additional circuit connections and components required.

Low-voltage-drop (0.6-V) regulators (such as the LM 2931) are suggested for battery operation. Other desirable features are invulnerability to battery polarity reversal, low quiescent current, and external turn-off.

21-8   Single-to-Dual Supplies

Occasionally only a single supply is available to power a circuit which normally requires a dual supply. An example is an op-amp powered by an automobile battery. One solution is to split the power-supply voltage in half, each half acting as a section of the dual supply (Fig. 21-16). In this case the power-supply tap is the same as the dual-supply common, which is normally connected to the circuit ground. If the battery is grounded, obviously the circuit cannot be also, but the common can be thought of as a floating ground for convenience in circuit analysis. If little current flows in the common return, i.e., the currents in the positive $+V_S$ and negative $-V_S$ supplies are nearly equal, only a center-tapped resistor would be needed to split the supply. Usually, however, the current drains from the negative and positive supplies differ substantially, and the supply slitter must be able to take up the difference. A unity-gain amplifier (Fig. 21-16a) works well for this purpose at modest current levels. If larger unbalanced (common) currents occur than the op-amp can handle (10 mA typically), an output dual-transistor current booster (Chap. 22) may be added. Voltage regulators may be added to one side of the dual supply as discussed above if the I/O voltage difference is sufficiently high (2 to 3 V).

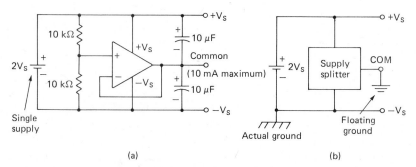

(a)   (b)

Figure 21-16   Split single supply as light-duty supply. The commons or grounds of the two suppliers differ.

21-9 DC-to-DC Converters

A dc-to-dc converter is a power supply with a dc input and dc output. The output is either higher in voltage than the input or of opposite polarity. Often the input is a battery. These converters generate a square wave which is rectified and filtered after passing through a capacitor or step-up transformer. In Fig. 21-17 two simple, relatively low-current converters based on the 555 timer are shown. Analysis of the diode rectification process has been described in Sec. 14-1. In Fig. 21-17a the output voltage is of the opposite polarity. This converts a single battery into a dual-polarity supply, as may be needed for op-amp circuits. In Fig. 21-17b a circuit modification is shown where the rectified (half-wave) and filtered voltage is added to the existing supply to produce an approximate voltage doubling. Because of the diode drops and ripple the voltage actually will be somewhat less. The

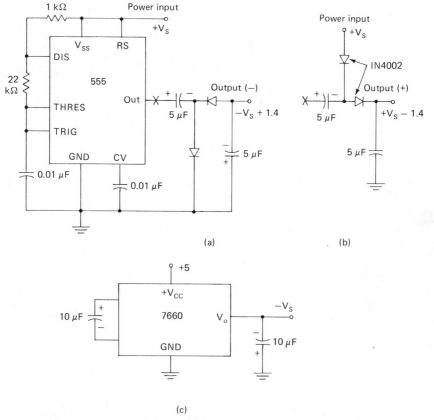

(a)

(b)

(c)

Figure 21-17 Lower-current dc-to-dc supplies: (a) voltage doubler; (b) negative output; (c) IC regulator version.

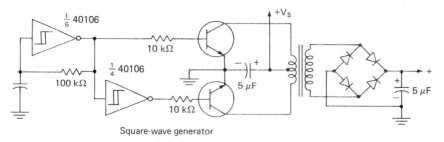

Figure 21-18 Transformer-coupled dc-to-dc converter power supply.

current capacity is somewhat low (30 mA) because of the IC limitations.

A specialized dc-to-dc converter IC which operates on the same principle as the 555 timer circuit is shown in Fig. 21-17c. It has a higher efficiency and has a voltage regulator (referenced to the input voltage) built in.

Where higher output current and voltage are required, a transformer driven by power-switching transistors can be used. A circuit utilizing widely available components is shown in Fig. 21-18. It operates at a frequency which is easy to generate and filter (5 to 50 kHz). Rectifier design considerations have been previously discussed. The square wave switches the two power transistors in a push-pull configuration. The output voltage on the capacitor is $2NV_S - 1.4$, assuming negligible ripple, where N is the transformer turns ratio (full primary to secondary). For example, if the transformer is a standard 12.6-V center-tape filament transformer (rated 110-V, 60-Hz primary) which has a turn ratio of about 9, the output voltage will be about 90 V when connected as a step-up transformer. Since the transformer is operated other than intended (higher frequency, primary and secondary reversed), the performance is difficult to predict accurately and some experimentation may be required.

Transformers which match specific transistors with specific input and output voltages are manufactured but are not widely available. They have the advantages of higher efficiency and simpler driving circuits, often because of an additional feedback winding. If a suitable transformer (and circuit) is available, it may be preferable to the more general circuit of Fig. 21-18.

21-10 Switching Regulators

In a switching regulator, the pass transistor is turned full on and off rapidly as a method of control (Fig. 21-19). Modulation of the switching pulse width, or more specifically, the ratio of the pulse on to off time

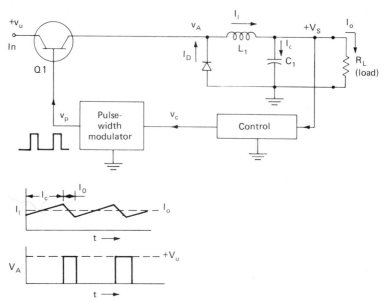

Figure 21-19   Principle of switching regulator.

adjusts the average current I_o and voltage $+V_s$ at the regular output. Efficiency is high and heat production low because the power dissipation in the switching transistor is small ($I_1 V_{\text{sat}}$ when on, or zero when off). For higher-power regulators the elimination of the heat sink (or reduction in size) is a substantial benefit.

An inductor (L_1 of Fig. 21-19) is required to smooth or average the current over the pulse period. When the transistor is on, the current through the inductor I_L increases and the capacitor is charged. When the transistor is off, the inductor current drops but does not change direction (assuming L is sufficiently large). During this part of the cycle, current flows through the diode D_1 and the polarity of the voltage across the inductor reverses. The switching frequency is made relatively high (10 to 100 kHz) in order to minimize inductor and capacitor sizes. The LC network is a low-pass filter which removes the switching frequency components but allows dc to pass.

A practical switching regulator is shown in Fig. 21-20. If the inductor has a magnetic core, it must not saturate at the rated current. A fast (Shottky) rectifying diode is needed at these higher frequencies. Transmission of the switching frequency interference back to the power line is avoided by the noise filter. It also reduces the coupling of line noise to the regulated output. Normally, the regulated output voltage is less than the input voltage, but it is possible to step up the

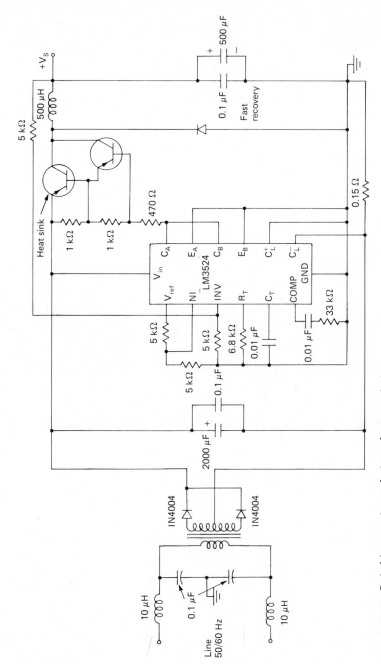

Figure 21-20 Switching power supply (step-down version).

voltage instead by connecting the inductor as a shunt rather than in series as shown.

21-11   Batteries and Battery Chargers

Portable battery-powered instruments are increasingly popular as a consequence of the development of smaller, low-power ICs. A list of widely available nonrechargeable batteries is given in Table 21-1. In general, the capacity of a particular type, expressed in amperehours (Ah), increases with battery weight at a given voltage. It is therefore possible to estimate the capacity of a battery by comparing its weight (or volume) with those of the same type given in the table, taking into account the number of cells. It should be observed that while the capacity differences between equal-sized batteries of different types is not extremely large, lithium and silver oxide cells are smaller for a given capacity and are best suited for miniaturized instruments. The shelf life of batteries is usually is over 1 year at room temperature and much longer at lower temperatures (refrigerated). Lithium batteries have a especially long shelf life (over 5 years). Capacity varies with manufacturing technique, shelf life, and desired endpoint voltage; the values in the table should be considered only estimates.

Battery capacity depends also on current magnitude (and cycling if intermittent). It will decrease by roughly 15 percent for each factor-of-2 increase in current beyond the suggested maximum. A carbon-zinc (Leclanche) cell has a voltage variation of 20 to 50 percent over its useful life. A mercury cell voltage will decline much less, about 2 to 10

TABLE 21-1   Nonrechargeable Battery Characteristics

Class	Type or size	No. of cells	Voltage	Capacity, Ah	Recommended maximum current, mA	Weight, g	Popular use
Carbon-zinc	AA	1	1.5	0.25–1	80	17	Penlight
	C	1	1.5	0.5–2	80	39	Flashlight
	F	4	6	2–10	250	640	Lantern
	2U6*	6	9	0.05–0.3	15	33	Transistor radio
Alkaline	VS1325[†]	6	9	0.1–0.5	30	40	Replacement for 2U6
Mercury	VSI[†]	1	1.4	0.3–0.5	80	12	Transistor devices
	M25	6	8.4	0.2–0.5	30	46	Transistor radio
Silver oxide	S5	1	1.6	0.07	3	1.2	Hearing aid
Lithium	1/2A	1	3.2	0.75	20	10	CMOS devices

* RCA number.
[†] Burgess number.

TABLE 21-2   Rechargeable Battery Characteristics

Class	Type or size	No. of cells	Voltage, V Nominal	Voltage, V Charging	Capacity, Ah	Trickle charge rate, mA	Recommended maximum current, mA	Popular use
Nickel-cadmium	(AA)	1	1.2	1.40	0.8	5	400	Type AA replacement
	CD25*	8	9.6	11.2	0.2	2	65	Transistor devices
Alkaline	C	1	1.5	1.75	1	25	200	Type C replacement
Gel	D	1	2	2.35	2.5	5	500	Portable instruments
	X	3	6	7.05	5	10	2×10^3	Portable instruments
Lead (oxide)	—	6	12	14.4	50	200	10×10^3	Automobile

* Burgess number.

percent until close to the end of its life, when the voltage drops rapidly. Alkaline (zinc–magnanese oxide) cells are similar to the carbon-zinc cell, but have about 50 percent greater capacity and can deliver more current without a loss in capacity.

No high-capacity, nonrechargeable 5-V batteries are readily available. A substitute is a 6.3-V lantern battery with two forward-conducting diodes in series to drop the voltage to 5.0 ± 0.5 V. If a 5-V regulated supply is desired, a low-drop (0.6-V) regulator is necessary when used with 6-V batteries.

Most of the comments concerning nonrechargeable cells, in particular the relation of capacity to weight, also apply to rechargeable cells (Table 21-2). Battery life, expressed as the number of times a battery can be charged and discharged, is maximized by keeping the current drain within the suggested maximum and avoiding complete discharge. However, battery exercise consisting of cycling from full charge to 90 percent discharge is beneficial.

Battery charging may be done at a slow rate (trickle charge at a roughly constant current), which does not require a voltage control, or at a fast rate, which does. A cell can be damaged (or even made to explode) by a high current, especially when fully charged; therefore either the charging current must be low or the state of charge, as measured by the terminal voltage, must be monitored.

Two trickle-charge circuits are shown in Fig. 21-21. Charging current I_C as set by the series resistor $I_C = (V_X - V_B)/R_S$ must not exceed the specified trickle-charge current when the battery is

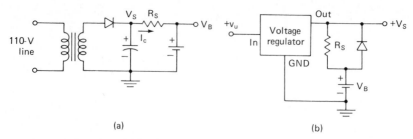

(a)   (b)

Figure 21-21   Battery trickle-charge circuits: (*a*) standard; (*b*) battery standby.

charged. In Fig. 21-21*b*, the battery is attached to the normal supply as a standby source of power in case of line power-supply failure. Although the voltage will drop from its normal regulated value of V_S down to V_B, many circuits will operate satisfactorily at a lower voltage, or at least stored digital data will be preserved.

Fast chargers (Fig. 21-22) simply consist of a current-limited power supply which turns on when the battery voltage drops below its fully charged value. The current limit of the regulator must be set at or below the maximum charging current to avoid battery damage. If several batteries are involved, they must be in series so that they charge and discharge together. Care must be taken to set the endpoint voltage to the battery's fully charged value accurately. If the endpoint is set too high, the charging current will not reduce to its trickle value when the battery is fully charged, while if it is set too low, the battery will not charge fully.

Battery life can be extended if the regulator voltage is reduced slightly (5 percent) at the endpoint voltage. When the battery voltage drops still lower, the regulator will switch again to its higher charging value.

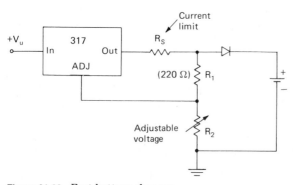

Figure 21-22   Fast battery charger.

22

Power Amplification
and Control Circuits

Occasionally an electronic instrument must deliver a relatively high power into an external load. Standard op-amps and ICs are low-power devices which are rather limited in the current and voltage they can deliver. Usually a power amplifier as a final stage provides the solution. Several types are described here. Also considered in this chapter are triac and SCR controls for loads such as motors or heaters placed across the power line.

22-1   Power Output Transistor Configurations

Power dissipation is an important and often limiting factor in power amplifier design. Discrete power transistors can dissipate considerably more power than ICs, especially when they are attached to a heat sink; for this reason they are often used in high-power devices. Efficient configurations are desirable both to increase the maximum available output power for a given set of transistors, to eliminate unnecessary drain from the power supply, and to reduce heating. Dissipation is minimized by keeping the power transistor current close to zero when no output voltage is required. With the push-pull configurations (class B or AB amplifiers) two transistors are required. One transistor controls the positive voltage output swing and the other the negative swing. Two types of configurations are shown in simplified form in Fig. 22-1. Bias is adjusted so that both transistors are slightly conducting when the input signal V_i is zero. On positive signal swings the current through Q1, and therefore the load R_L, increases while Q2 turns off. On negative swings Q2 is on and Q1 is off. The bias diodes compensate for

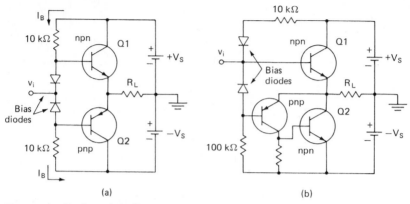

Figure 22-1 Basic push-pull output configurations with dc response: (*a*) complementary symmetry; (*b*) totem pole with driver.

the transistor base-emitter voltage. When properly adjusted, collector currents (Q1 and Q2) are small (1 percent of peak value) with zero signal. Some means of bias equalization between transistors is desirable, e.g., by the addition of small emitter resistances. Sometimes separate driving stages are provided for Q1 and Q2. A close match between transistors is difficult to achieve, especially at the crossover point where one transistor is turning on while the other is turning off. As a consequence of this mismatch, a notch is often observed in the signal near zero.

Care must be taken to avoid turning Q1 and Q2 on simultaneously, as might occur with improper bias, since in effect the power supplies are then shorted through the transistors. When hot, transistors draw additional bias current, especially when biased from a fixed voltage source. Transistor power dissipation is the average (over 10 to 100 s) of $I_c V_{ec}$, where I_c is the collector current and V_{ec} is the voltage drop across the transistor. Under some conditions the increase in power dissipation with bias current can heat the transistors enough for the transistors to be destroyed, a condition known as thermal runaway. Current limit of some type is essential.

Where efficiency or power dissipation is unimportant, the simple transistor emitter-follower configuration may suffice (Fig. 22-2). The emitter resistor R_E determines both the maximum (negative for *npn*) signal current I_n and the zero-signal transistor dissipation P_D. Current limit is provided by R_s. Dual-polarity high-current power supplies are needed with the configurations shown, a disadvantage since a high current is required. For ac output, a single power supply can be used if the load is coupled through a capacitor (and the input signal is properly biased). Practical versions of this circuit are described below.

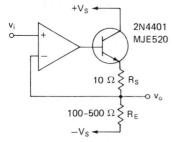

Figure 22-2 Op-amp with emitter-follower current booster.

Another type of power amplifier which utilizes switching transistors is described in Chap. 21.

22-2 High-Output Op-Amps

Higher current can be drawn from the output of an op-amp booster. The booster of Fig. 22-3 utilizes the complementary symmetry configuration. Notice that the transistor amplifier output V_o is the effective op-amp output (rather than V_o') and any op-amp feedback configuration may be used. The emitter resistors R_E limit the output current as well as provide bias equilization between transistors. Feedback reduces the crossover notch to a negligible level, but with some transistors a high-frequency oscillation may occur near the crossover point as a consequence of the larger phase shift or poorer frequency response near transistor cutoff. This problem usually can be eliminated by the addition of a feedback capacitor C_f or application of another op-amp feedback-compensation method.

High-power op-amps are available for high-current or high-voltage

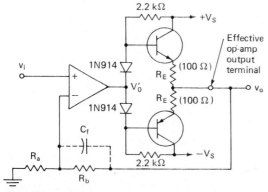

Figure 22-3 Op-amp output-current booster. Noninverting configuration is illustrated.

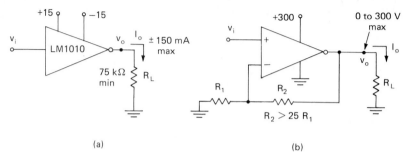

(a)   (b)

Figure 22-4   Power output op-amps: (*a*) high-current buffer and (*b*) high-voltage op-amp.

loads. For higher currents, a unity voltage gain buffer (Fig. 22-4*a*) may be satisfactory. It has the same output voltage as a normal op-amp but has ten times the current. For higher voltages, a special op-amp which has about the same current output as a normal op-amp but has 10 times the voltage is available (Fig. 22-4*b*).

22-3   AC Amplifiers

While the amplifiers described above can amplify alternating as well as direct current, they have the inconvenience of requiring a high-current negative supply as well as a positive supply. The ac amplifiers described have only a single supply. Transformer or capacitor output coupling is involved. Where distortion is not critical, a transformer-coupled (class B) push-pull amplifier (Fig. 22-5) may be suitable. Here Q1 conducts on the positive signal cycles and Q2 on the negative. One driver (IC1) and diode provides the proper bias while the other driver

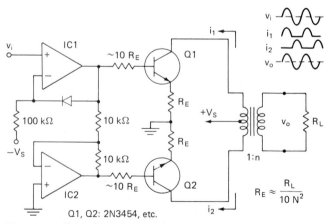

Figure 22-5   Transformer-coupled ac amplifier.

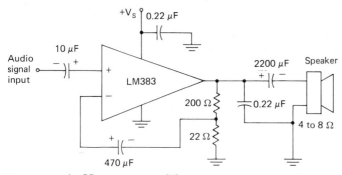

Figure 22-6 An IC power ac amplifier.

(IC2) inverts the phase for Q2. The transformer combines the transistor currents and steps the voltage up (or down). If we take into account the voltage drop across R_e, the peak-to-peak output V_o is approximately $0.9V_sN$, where N is the turns ratio between the total primary and the secondary.

Complete IC ac amplifiers capable of delivering modest power into a low-impedance load R_L such as a 4-Ω speaker, have been developed for the consumer electronics industry. An example is given in Fig. 22-6. The low-frequency response is determined by the external capacitors, especially the output capacitors C_o. Internally most circuits are similar to those of Fig. 22-1. Normally a single, unregulated supply $+V_s$ is all that is required. Metal strips (tabs) which act as heat sinks often protrude from the IC package. If the strips are connected to larger heat sinks, this unit may deliver 1 to 5 W. Some packages contain two amplifiers which may be used separately (stereo), or together in a "bridge" configuration where the load is placed between the two amplifier outputs for greater peak-to-peak output voltage.

22-4   SCR and Triac Drivers

Both the SCR and triac are high-current high-voltage semiconductor switches, often termed *thyristors*. The SCR will pass current in one direction, but the triac will also pass alternating current. Applications of the SCR include high-voltage power supplies and dc motor control. Triacs are popular controls for loads operating from the 110- to 440-V ac lines.

An SCR is turned on by applying a voltage between the gate and cathode sufficient to raise the gate current to a threshold value (Fig. 22-7). When switched on, the SCR is equivalent to two forward-conducting diodes in series (voltage drop 1.2 to 1.6 V). It remains on until the cathode-to-anode voltage becomes less (more negative) than a

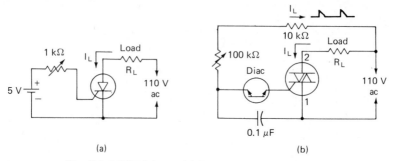

Figure 22-7 Simplified SCR drivers: (*a*) dc gate current; (*b*) diac (pulse) control.

holding voltage (about $+ 1.2$ V), which for an ac anode voltage occurs at the end of the positive half of the cycle. Gate-current requirements vary widely, ranging from below 1 mA for a "sensitive-gate" SCR to well over 100 mA for a higher-anode-current (5- to 50-A) SCR. Many require a higher gate current than can be delivered by the usual logic or op-amp ICs. However, the gate current need be applied only briefly (typically 2 to 20 μs), and therefore most SCR drivers are designed to deliver short pulses which have a high peak but a low average current (to reduce power requirements).

A popular SCR-gate pulse generator, when the control is derived from a higher-voltage source, is the diac or trigger diode (Fig. 22-7*b*). It has the property of remaining nonconducting (off) until a threshold voltage is reached (typically 20 to 40 V), at which point it suddenly becomes highly conducting until the voltage drops to a low value (<1.2 V), when it again switches off. Almost invariably the diac is connected between a charged capacitor and the gate so that when the capacitor voltage reaches the diac threshold value, it discharges through the SCR gate, producing a high peak current.

Gate current requirements for the triac are similar to those of the SCR if the gate polarity is proper and/or the triac is of the proper type. Terminals 1 and 2 of triacs correspond roughly to the cathode and anode of an SCR, respectively. Under conditions of ac gate drive, where the gate-to-terminal-1 voltage and terminal-2-to-terminal-1 voltage have the same polarity (quadrants I and III), the gate turn-on current (magnitude) is comparable to that of the SCR. In the circuit of Fig. 22-7*b*, the triac is substituted for the SCR since the diac driver will deliver a trigger pulse during both the positive and negative cycles. Most triacs can also be triggered by a positive gate voltage when terminal 2 is negative (quadrant IV operation). In this case any drive circuit which works for the SCR will work for the triac. Care must be taken to determine whether a triac will operate in quadrant IV if a positive drive is chosen, since some types of triacs have an extremely

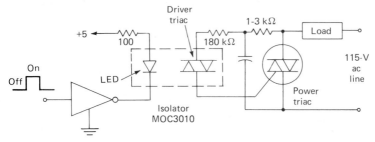

Figure 22-8 Isolated optical isolator triac driver.

poor sensitivity or can be damaged if operated in this quadrant. By contrast, other types of triacs may require such a low gate current that they can be turned on by a CMOS driver.

Usually triacs control loads such as motors or heaters which are connected directly to the power line. Since neither side of the power line (hot or neutral) may be connected to circuit ground, a ground-isolated drive circuit is required. Optical isolators are best, although pulse transformers are an acceptable alternative. For power triac drive applications, a special isolator which combines a photodiode and a low-current driver-type triac (Fig. 22-8) is best. The driver-triac is thus light-activated so that when the LED is turned on, the triac also goes on. The driver triac then turns on the power triac by providing a gate current pulse I_g. When the main triac is on, the voltage to the driver triac becomes small as does I_g. Although the driver triac can supply only limited current, it is rated to withstand the peak line voltage. Control of power triacs which can switch loads of several kilowatts are possible.

22-5 Small Motors and Drivers

Switching or speed control of small motors is occasionally required for instrumentation applications. The types of motors likely to be encountered are reversible dc motors, reversible ac motors, three-phase servometers, synchronous ac motors, and stepping motors. A dc motor with a permanent magnetic field, e.g., the small motors used in toys, can be driven with a range of dc voltages. Direction of rotation reverses with polarity reversal. Speed increases with voltage and decreases with load. Speed is not constant without a velocity sensor (tachometer) and feedback control. A power amplifier with a dc output is a suitable driver. The magnitude of the supply voltage V_s should be equal to the maximum or full-speed voltage. Of course, the power supply and power amplifier must be able to deliver the current required by the motor under full-load conditions. It is desirable to set the power-supply

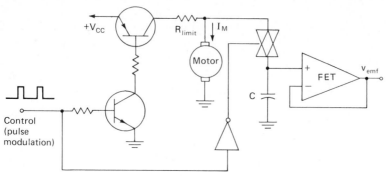

Figure 22-9 Pulse-modulated small motor control with back-EMF speed pickup.

current limit just beyond the full-load current point to provide protection in case the motor stalls and draws excess current. Control of motor speed and direction is accomplished by varying the amplifier input voltage and polarity.

The back electromotive force produced by the motor V_{emf} may be used to determine motor speed. It must be measured with the motor current excitation I_M off as is done in the control of Fig. 22-9. Since motor speed is easily controlled by pulse modulation (width or frequency), a track-and-hold may be activated during the period when the motor current is zero.

A common reversible ac servomotor has separate field and control windings. The field winding is connected to the line (usually 110 V, 60 Hz). The control winding is connected to an ac source which is phase-reversible and of variable amplitude. The motor speed increases with control voltage up to its maximum, where the control voltage is equal to the line voltage. Motor direction reversal is accomplished by reversing the control-winding phase (from approximately 0 to 180° with respect to the field-winding phase). A transformer-coupled ac power amplifier (see Fig. 22-5) is a suitable driver.

Three-phase servomotors are sometimes employed for remote positioning, especially where relatively rapid response is required. In principle, and occasionally in practice, the connection between the transmitter and receiver servos is direct (Fig. 22-10). The transmitting servo can be thought of as a transformer with a primary (rotor) which can be rotated continuously over 0 to 360°. There are three secondaries (stators) mechanically fixed in position at 120° angles. As the rotor is moved, the amplitude and phase of the voltages induced in the three windings vary and are a unique function of angular orientation. The receiver servo has the same construction, and the stator voltages are

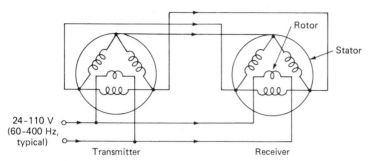

Figure 22-10 Three-phase servo connection.

the same as the transmitter. Correspondingly the magnetic fields in the receiver match the transmitter at any instant, and therefore the receiver rotor moves to match the position of the transmitter. More powerful servos require an ac amplifier on each phase. Actually it is the winding voltage phase angles which determine the receiver position, and the transmitter and a servo can be replaced with a circuit which generates signals with the appropriate phases.

The speed of synchronous motors is locked to the line frequency. Since the line frequency (60 Hz) is constant, especially over the long run, they are good clock motors. Control of speed over a modest range at reduced output torque can be accomplished by a variable-frequency drive.

22-6 Stepping Motor Drivers

A stepping motor is a pulse-controlled motor which rotates in discrete steps, typically 100 or 200 per revolution. One step of a 100-step motor, for example, corresponds to an angular rotation of 3.6°. It has a reproducible angular position, which is desirable in many instrumentation applications, although the speed and torque are mediocre.

The operating principle of the stepping motor is shown in Fig. 22-11 for a hypothetical four-step motor. When no coils are energized, the permanent-magnet rotor remains fixed in position because it is attracted to, and held by, the closest stator pole. It can be moved to a neighboring pole by applying sufficient current through the electromagnetic coil of that pole. Suppose, for example, that the north pole is at position D. If pole A is now magnetized as a south pole, the rotor north pole will swing to A. It can be seen that if the poles are pulsed in an $ABCD$ sequence, the rotor will continuously rotate clockwise, and if pulsed in an $ADCB$ sequence, the rotation will be counterclockwise. Practical stepping motors have many more poles on the stator (and

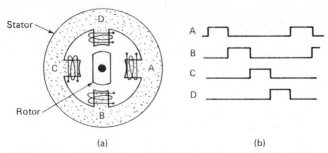

Figure 22-11 Stepping-motor principle: (*a*) four-step motor (hypo-thetical); (*b*) winding step sequence (same for 100- and 200-step motors).

rotor), but they are often organized in sets of four, so that the same sequence of four pulses into four windings is required. The motion can be stopped or reversed after any step.

A practical drive circuit for a stepping motor is given in Fig. 22-12. The step sequence given is controlled by a 2-bit counter and decoder. Greater motor torque is obtained by overlapping the pulses and doubling their pulse length. Each input pulse produces one step. The direction of the counter and motor is controlled by the counter up-down control.

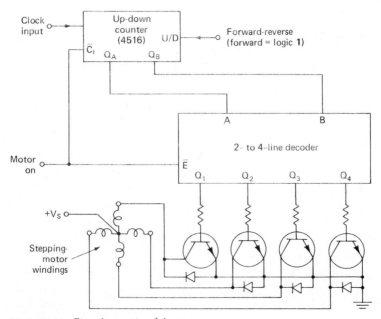

Figure 22-12 Stepping motor driver.

In order to conserve power a gate connected to the transistor drivers turns off the motor current during prolonged periods where no movement is required.

POWER CIRCUIT DESIGN EXAMPLE

Example 7   Dual 15-V Supply

DESCRIPTION: This standard +15/−15-V regulated supply will deliver 70 to 150 mA and is suited for powering op-amp circuits.

SPECIFICATIONS
   Power line: 115 V, 60 Hz (15 W)
   Output voltages: ±15.0 ±0.3 V
   Regulation (line or load):   1 percent
   Maximum current: 70 mA (average, both sides)
                                         120 mA (short-term)
   Protection: short-circuit and thermal overload

DESIGN CONSIDERATIONS: Two standard three-terminal IC regulators with internal current and thermal limits were selected. The input voltage must be 3 to 5 V higher than the regulated output, so that the unregulated voltages should be +19 and −19 V. Allowing for diode drop and assuming a 10 percent ripple, the transformer should deliver 20.7 V peak (14.7 V rms) minimum across each half secondary. A 30-V rms center-tapped transformer was selected. From Fig. 22-4 the input capacitors should be 350 μF or more for a 120-mA load (10 percent ripple). Power dissipation for a 120-mA load is 1 W.

POWER CIRCUIT DESIGN PROBLEMS

37   Design a regulated power supply that will deliver ±12 V (dual) at 100 mA and +5 V at 200 mA using only one transformer (26-V rms center-tapped at 1A). Calculate the power dissipation in the regulators and choose a proper heat sink if needed. A resistor in series with the 5-V regulator (chosen so the regulator input voltage is 8 V at full load) will minimize heating.

38   A power supply for a microprocessor requires a battery backup. The battery is 6.6 V at full charge (3-Ah gel-type). Voltages and currents required

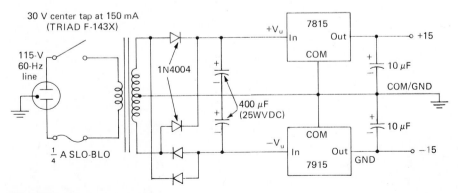

Figure E-7 Dual power supply.

are $+5$ V at 50 mA, $+15$ V at 20 mA, and -15 V at -10 mA. Regulation is not necessary (these are the specifications of the S-100 bus many microprocessors use). Design such a supply, including the battery charger.

39 Design a triac circuit that will control the speed of a motor (110 V, 10 A maximum) from a microprocessor port. Isolation of the power line from the port ground is necessary.

40 Design a power amplifier (voltage gain of 10) that will deliver an output (ac or dc) of ±30 V at 1 A (30 W peak). A "high-voltage" op-amp and discrete transistors are suggested. For more of a challenge, redesign for an output of ±100 V maximum.

Index

ABOUT THE AUTHOR

Darold Wobschall is an Associate Professor of Electrical and Computer Engineering at the State University of New York at Buffalo, where he teaches electronic instrument design, and is president of Index Electronics, Inc., where he practices electronic instrument design. He has published dozens of articles in scientific journals, many having to do with new instruments or experimental techniques, and he also conducts seminars for professional engineers.